Frederick Bateman

**Darwinism Tested by Language**

Frederick Bateman

**Darwinism Tested by Language**

ISBN/EAN: 9783743394902

Manufactured in Europe, USA, Canada, Australia, Japa

Cover: Foto ©berggeist007 / pixelio.de

Manufactured and distributed by brebook publishing software (www.brebook.com)

Frederick Bateman

**Darwinism Tested by Language**

# DARWINISM

# TESTED BY LANGUAGE

BY

FREDERIC BATEMAN, M.D.,

FELLOW OF THE ROYAL COLLEGE OF PHYSICIANS;
PHYSICIAN TO THE NORFOLK AND NORWICH HOSPITAL;
CONSULTING PHYSICIAN TO THE EASTERN COUNTIES' ASYLUM FOR IDIOTS;
FOREIGN ASSOCIATE OF THE MEDICO-PSYCHOLOGICAL SOCIETY OF PARIS;
AUTHOR OF "APHASIA AND THE LOCALISATION OF THE FACULTY OF LANGUAGE."

WITH A PREFACE BY

EDWARD MEYRICK GOULBURN, D.D.,

DEAN OF NORWICH.

RIVINGTONS,
*WATERLOO PLACE, LONDON,*
𝔒𝔵𝔣𝔬𝔯𝔡 𝔞𝔫𝔡 𝔆𝔞𝔪𝔟𝔯𝔦𝔡𝔤𝔢.
STACY,
2, *HAYMARKET, NORWICH.*

MDCCCLXXVII.

# PREFACE.

There are two contrary intellectual tendencies, which characterize minds of different orders, and, when indulged to excess, become intellectual vices. The one is the tendency to see a distinction where there is no real difference. This is the snare of cultivated (or perhaps of over-cultivated) minds, whose constitution may never have been robust, and what vigour they once had has been refined away by speculation. To see a distinction without a difference is the vice of the trained and subtle thinker. Opposed to this is the tendency to ignore real differences; to bring rapidly under the same category two cases which have one or more superficial features of resemblance, but which are so fundamentally unlike that they cannot with any justice be classed together. It may have often happened to us to meet with a stranger, who has some one common feature with a person of our acquaintance. In virtue of his having such a feature he reminds us for a moment of that person; but, when we take a second look, we see that the resemblance is only on the surface; the whole head and bust are of a different type altogether. But in matters intellectual, a resemblance sometimes seems so captivating (especially if our own researches have brought it to light), that we do not take the trouble to look at the plain and deep-seated differences, but treat it as a real analogy, and rest the weight of a whole theory upon it. It must be, one would think, under the power of some hallucination of this kind, that the disciples of Evolution venture to deny the existence in man of a new and distinguishing element, over and above the nature which he has in common with the lower animals. How this distinction can be matter of doubt to any one, except under the fascination of a favourite theory which blinds the mind to every thing subversive of itself, is truly surprising. The prerogative of man is not an assertion of theology merely. It is written not more clearly on the pages of the Bible than on the common sense and experience of all the world. There seems to be a wide gulf even between vegetable life and brute matter; a wider still between the sensibility and instinct of animals and vegetable life; and a gulf perfectly impassable between this sensibility and instinct, and the conscience, reason, and capability of civilization, which we find in man.

We need not deny or undervalue the discovery that certain higher and more advanced forms of vegetable and animal life developed themselves originally out of lower and more rudimentary forms, according to certain laws supposed to be ascertained by Mr. Darwin and others— struggle for existence, survival of the fittest, developement of resources under pressure of necessity, &c., &c.; but, carry back the series as far as you will, must not the earliest germ of vegetable and still more of animal life have been a new introduction into the system, which nothing that existed previously could have given rise to? Out of a piece of ore, out of a clod of earth, can you generate life? And when we look at man, the differences that part him off from the lower animal creation are so trenchant and so significant, that one would think that those philosophers, who maintain that he is merely an animal, with its powers developed to the highest degree, can never have looked them full in the face, under the conviction that to do so would disturb their theory. These differences may be briefly stated as three. Man can speak; he can make improvements in his own condition, to which it is difficult to set limits; and he can worship.

The first (and perhaps the most fundamental) of these differences Dr. Bateman has exhibited very ably and pointedly in the work which is now presented to the reader. He aims at illustrating the truth in "the grand old book," that "God made man in his own image; in the image of God created he him;" and with this view he shows that (just as in the precinct of the Divine Nature the Word, or Second Person, represents the Father, and reveals the Father to the creatures, so) the word is man's distinguishing characteristic, represents him, is the great medium whereby he throws into other minds the thoughts conceived in his own. Language is unquestionably the great outcome of Reason; indeed it *is* the Reason, not indeed ἐνδιάθετος, (viewed as latent in the mind), but προφορικὸς, (expressing itself outwardly). Let it be considered how much classification there is even in the humblest sorts of language; how the mere use of an appellative, like *gate, book, field,* to denote a whole class of objects, is the result of a classification, in order to arrive at which individual differences must be overlooked, and a general idea formed in the mind; how epithets denote qualities, and the idea of qualities is formed by the mental power of abstraction, which strips off from several objects some particular feature in which they agree — let this be considered, and it will be seen at once that Language is a popular philosophy, and surely (as such) entirely out of the reach of the lower animals, the most sagacious of which can never be supposed competent to such mental processes as abstraction and generalization. Dr. Bateman shows, by describing an interesting case which came under his own notice at Paris, (P. 108) that mere phonetic mimicry is not language; there is no mind in it; it is a trick of the ear. The evidence which he has amassed and advances to show that all men have the faculty of language (at least in the germ), and that no creatures but men have, seems to be thoroughly satisfactory and conclusive.

The present work being rather of a scientific than a general character, the author has chiefly exhibited the Reason in its most primary and pure operation, as giving birth to language, and has not gone on to consider it in its application to the life of man, and in the various reliefs of his present condition which it affords. This is the second difference which parts us off from the lower animals; and it is a difference quite capable of being appreciated by the most unscientific of minds. Brutes have never made the smallest approach towards civilization. Of arts, whether useful or ornamental, not a trace has been ever found among them. Never have they been known to manifest a single glimmer of that faculty by which one generation of mankind takes up the discoveries and researches of its forefathers, and makes them the basis of a material advance in the arts of life, and a stage in human progress. This was what the subtle and profound Blaise Pascal pointed out so well long ago; and assuredly no subsequent experience of men or animals has obliterated the distinction which he expounds so luminously.

"N'est-ce pas là *traiter indignement la raison de l'homme, et la mettre en parallèle avec l'instinct des animaux*, puisqu'on en ôte la principale différence, qui consiste en ce que les effets du raisonnement augmentent sans cesse : au lieu que l'instinct demeure toujours dans un état égal ? Les ruches des abeilles étaient aussi bien mesurées il y a mille ans qu'aujourd'hui, et chacune d'elles forme cet hexagone aussi exactement la première fois que la dernière. Il en est de même de tout ce que les animaux produisent par ce mouvement occulte. La nature les instruit à mesure que la nécessité les presse ; mais cette science fragile se perd avec les besoins qu'ils en ont : comme ils la reçoivent sans étude, ils n'ont pas le bonheur de la conserver ; et toutes les fois qu'elle leur est donnée, elle leur est nouvelle, puisque la nature n'ayant pour objet que de maintenir les animaux dans un ordre de perfection bornée, elle leur inspire cette science simplement nécessaire et toujours égale, de peur qu'ils ne tombent dans le dépérissement, et ne permet pas qu'ils y ajoutent, de peur qu'ils ne passent les limites qu'elle leur a prescrites.

Il n'en est pas ainsi de l'homme, qui n'est produit que pour l'infinité. Il est dans l'ignorance au premier âge de sa vie ; mais il s'instruit sans cesse dans son progrès : car il tire avantage non seulement de sa propre expérience, mais encore de celle de ses prédécesseurs ; parce qu' il garde toujours dans sa mémoire les connaissances qu'il s'est une fois acquises, et que celles des anciens lui sont toujours présentes dans les livres qu'ils en ont laissés. Et comme il conserve ces connaissances, il peut aussi les augmenter facilement ; de sorte que les hommes sont aujourd'hui en quelque sorte dans le même état où se trouveraient ces anciens philosophes, s'ils pouvaient avoir vieilli jusqu' à présent, en ajoutant aux connaissances qu'ils avaient, celles que leurs études auraient pu leur acquérir à la faveur de tant de siècles. De là vient que *par une prérogative particulière*, non-seulement chacun des hommes s'avance de jour en jour dans les sciences, mais que tous les hommes ensemble y font un continuel progrès, à mesure que l'univers vieillit, parce que la même chose arrive dans la succession des hommes que dans les âges différents d'un particulier. De sorte que toute la suite des hommes, pendant le cours de tant de siècles, doit être considérée comme un même homme qui subsiste toujours, et qui apprend continuellement."

This noble passage, while it is an utterance of one of the most acute and philosophical minds which ever existed, is not less the dictate of common sense; and it is to be wished that our modern men of science would lay to heart what so luminous and profound a writer says, as to its being *an unworthy treatment of human reason to put it on a level with the instinct of animals*, and as to man's corporate inheritance of the treasures of knowledge being a *prérogative particulière* of our race.

But the third obvious and fundamental distinction between man and the inferior animals consists in the conscience or religious instinct. Holy Scripture, in enumerating the different parts of our nature, distinguishes between the spirit and soul of man (1 Thes. v. 23), and shows that this distinction is a real one, and that the two words are not used together as a periphrasis for the immortal part of man, by speaking in sharp contrast of the man of the soul ($\psi\upsilon\chi\iota\kappa\grave{o}s$) and the man of the spirit ($\pi\nu\epsilon\upsilon\mu\alpha\tau\iota\kappa\grave{o}s$),

(1 Cor. ii. 14, 15); of the body which is the organ of the soul (σῶμα ψυχικόν) and the body (to be brought into existence at the Resurrection) which shall be the organ of the spirit (σῶμα πνευματικόν), (1 Cor. xv. 44). We shall not probably err much if we regard the soul (in this precise and accurate sense of the term) as, no less than the body, attaching to animals in common with man. Many animals manifest in a remarkable degree that lower species of intelligence to which Coleridge, following in the wake of the German philosophers, gives the name of understanding as distinct from reason; and they undoubtedly are sharers in many of the affections,—fear, emulation, pity, and the parental instinct,—all of which are seen in man also, but in him are dignified and raised to a loftier platform,

as being kneaded up in the same personality with the immortal spirit. This immortal spirit has two operations, the intellectual, (with its powers of induction and deduction), and the moral or devotional,—the one giving us assurance of the being of a God, the other recognising Him as Lord of our consciences, and leading us to yield Him worship. Where among animals is there the faintest glimpse of so sublime a faculty? Take the faculty in the utmost state of degradation and debasement, in which it has been ever found. Let the only things correspondent to the religion and worship of the highly civilised man be a superstitious regard to some fetish, supposed to exercise a power of blight over harvests, and over the lives of men and cattle, and a number of foolish (and perhaps bloody) rites designed to break the spell,—let the religious instinct be plunged as low as this in darkness and bondage,—yet where will the least parallel or approach to such sentiments and usages be found among the inferior animals? It may be conceived that even out of a faculty so debased, there might be produced, by pouring the light of Divine truth upon it, and raising the general civilization of the whole man, a conscience which should recognise the true God as its Judge, and the atonement of the true Saviour as its hope, and make itself the controlling principle of the entire moral life; but round what nucleus in the intelligence and feelings of *an animal* such sentiments could form, we must leave it to the professors of "Evolution" to explain. Ordinary minds are unequal to the task.

Dr. Bateman, arguing the question, as a man of science, chiefly on scientific grounds, has only briefly alluded to this branch of the subject. Still, as a Christian in something more than the name, he has felt that his Essay would be incomplete without the notice of the religious instinct, and has devoted to this *differentia* of man some of his ablest and most interesting remarks (Pp. 208—217). The reader will be of opinion that he has compressed into a very brief compass a large amount of weighty and telling argument. It is to be hoped his argument may have its effect against the crude theories and unsupported hypotheses of the disciples of Evolution. For assuredly it is something more than a mere speculative view of man's origin which is at stake. To degrade man in theory, to instil into him that he is no more than a superior and highly cultivated animal, and to obscure or throw out of sight his distinguishing relation to the Father of spirits, is the surest way to degrade him morally; and, should such teaching ever take a strong hold of the public mind and prevail extensively (which is hardly probable), would surely achieve that end. This self-degradation of man has been his tendency from the first beginning of his history. Idolatry made the primitive races of mankind bow down before the visible objects of Nature, before the creatures brought into existence to do them service, and even before the effigies of these. And now that a Christianized civilization has rendered this gross form of idolatry impossible, the tendency to self-debasement re-appears in the shape of a scientific speculation, the scope of which is to veil all man's higher affinities and instincts, and to throw into strong relief his affinities with the creatures below him, —a new and weird fulfilment of the old complaint lodged against God's people, THUS THEY TURNED THEIR GLORY INTO THE SIMILITUDE OF A CALF THAT EATETH HAY,— a fulfilment impressing that old lesson, which is one of the keys to History, that, while manners shift, and the fashion of this world passeth away, men are still, in their fundamental weaknesses and temptations, what they ever were.

E. M. GOULBURN.

*The Deanery, Norwich,*
 *August 1st,* 1877.

# CONTENTS.

## CHAPTER I.

Preliminary remarks — Darwinism defined—Man's Genealogical Tree—The "Missing Link" between Man and the Man-like Ape—Professor Haeckel's Natural History of Creation—Lemuria, the birth-place of the Ape-like Man—The Ascidian, Man's remote ancestor—Dean of Canterbury's reflections on the Ascidian descent of Man—Haeckel's Moner — The Regnum Protisticum or Kingdom of Primitive Forms. .......... 1

## CHAPTER II.

Evolution theory contrasted with the Scriptural account of the origin of Man—The Monistic and Dualistic hypotheses—The Primordial Germ—What brought it into existence?—Archdeacon Pratt's and Dr. Reichel's views—Mr. Disraeli on Evolution. .......... 41

## CHAPTER III.

Sentimental opposition deprecated — Broca, Max Müller—No evidence of the transmutation of species within the historic period—Flourens—Animal kingdom of Aristotle, the same as that of our day. Plea of the Imperfection of the Geological Record considered—Haeckel, Duke of Argyll, and Mivart—Professor Agassiz on the Immaterial Principle—The Brain of Man compared with that of the Orang, Chimpanzee, and Gorilla. .................... 55

## CHAPTER IV.

Definition of Language—Stoddart, Trench, Whitney, and Farrar — Language, a Distinctive Attribute of Man — The so-called language of the Parrot considered—Connection between Ideas and Words—Cases of Heterophasia or Perversion of the Faculty of Language. ............... 87

## CHAPTER V.

The Anatomical Seat of Speech—Rôle of the Cerebral Convolutions; Flourens, Maudsley—Gall's Phrenological System—Destruction of the anterior lobes of the brain without impairment of the power of speech—Comparative development of the third frontal convolution in Man and in the Ape—Speech is a barrier the brute is not destined to pass. 116

CHAPTER VI.

Language is a Distinctive Attribute of Man— Man *versus* Ape controversy — On the Universality of Language — Is there a Speechless Tribe?—The Fuegians and the Veddahs of Ceylon — Tylor, Lubbock, Whitney, and Trench — The so-called speechless wild Men were probably Apes— Evidence of the great travellers of the day —The Soko of Dr. Livingstone, and the Orang Outangs of Lord Monboddo. ...... 147

CHAPTER VII.

The Immateriality of the Faculty of Speech—The Brain a mere Instrument— The Electric Telegraph and its Language— Inconsistencies of the Evolutionists—The Odium Antitheologicum — The Modern Anthropological Creed—American Writers on Evolution — The Tripartite Nature of Man — The German Neologists on Life, Matter, and Force—The Mystery of Life— Conclusion. ......................... 178

# ILLUSTRATIONS.

|  | Page |
|---|---|
| The Man-like Ape | 9 |
| Catarrhine or Old World Monkey (Macacus) | 19 |
| The Ascidian. Our Pre-Historic Ancestor | 24 |
| The Moner. Man's First Ancestor | 31 |
| The Second Group of the Protistic Kingdom | 35 |
| A Flagellate | 37 |
| Brain of Man *(Homo)* | 84 |
| ——— Orang *(Simia)* | 84 |
| Convex Surface of the Left Hemisphere, showing the Disposition and Arrangement of the Cerebral Convolutions | 127 |

# CHAPTER I.

"To whom the winged hierarch replied:
O Adam, one Almighty is, from whom
All things proceed, and up to him return,
Endued with various forms, various degrees
Of substance, and, in things that live, of life."

PARADISE LOST.

*Preliminary Remarks—Darwinism defined—Man's Genealogical Tree—The "Missing Link" between Man and the Man-like Ape—No fossil remains of the Ape-like Man—The Ascidian, man's remote ancestor—Dean of Canterbury's reflections on the Ascidian descent of Man—Haeckel's Moner—The Protistic Kingdom.*

Perhaps no works in modern times have been so largely read and so freely criticised, and have exercised so great an influence for good or for evil, as the "Origin of Species" and the "Descent of Man."

B

The subject of which they treat is one of such absorbing personal interest, as tending to gratify the ardent desire for knowledge of the "*where, the whence, and the whither*," of the human race, that these books have been received and perused with avidity, not only by professed naturalists, theologians, and men of science, but by a far wider circle of general readers.

It has been said of Luther that he was the monk that shook the world. It may with equal propriety be said that Mr. Darwin is the naturalist, who, by a hypothesis so strangely at variance with our traditions, has threatened to shake the foundations of the religious world. The theory enunciated in his writings, trenching as it does upon questions of the last importance and of the most absorbing interest to man, has been welcomed by acclamation by some, anathematized by others; and so numerous have been the publications of the opposing parties, that a whole special literature may be said to have sprung up, having for its key-note, the Evolution Theory.

During the last few years, there has been an increasing desire, on the part of the earnest and thoughtful members of the community, to investigate apparent discrepancies between Christianity and Science, and to deal with some of the modern forms of supposed antagonism between Science and Scripture, and, as in my opinion, the Darwinian hypothesis of the origin of man is directly opposed to the teaching of Revealed Religion, I purpose making a calm, dispassionate, and unprejudiced inquiry into the value and truth of those doctrines as to man's relationship to the Simian families, which, during the last ten years, have acquired such a rapid, but, as I believe, undue, development amongst large classes of society both in Germany and England. The novelty of Mr. Darwin's views has had something to do with the ready reception of them by the rising generation, who, in this age of electric telegraphy and underground railroads, are always seeking the sensational and the marvellous, the tendency of whose mind is to consider those who differ from them as standing upon a lower intellectual platform than themselves.

It is not my intention to dwell at any length on the peculiar scientific views which we understand by the term, Darwinism, but, as I have reason to believe that there are still many persons who have but an imperfect idea of what the doctrine of evolution really means, I will very briefly give a definition of it.

In his work on the "Origin of Species by Natural Selection," Mr. Darwin promulgated the theory, which had been previously put forth by the French Zoologist, Lamarck,* that all species, instead of having been independently created, and possessing an independent existence, had been gradually developed out of other forms, and that all organic beings that have ever lived on this earth have probably descended from some one primordial form, into which life was first breathed by the Creator.† In this treatise he merely hinted

* Philosophie zoologique, ou exposition des considérations relatives à l'histoire naturelle des animaux. Paris, 1809.

† "I believe that animals have descended from at most only four or five progenitors, and plants from an equal or lesser number. Analogy would lead me one step

at the application of his hypothesis to man, remarking that in the distant future he saw open fields for far more important researches; that psychology would be based on a new foundation, and light would be thrown on the origin of man and his history, but in his recently published work, he accepts the responsibility of the application of his theory to the human race, to which he applies all the consequences of his doctrine; and he does not hesitate to assert that Man, the wonder and glory of the universe, is descended from a hairy quadruped, furnished with a tail and pointed ears, probably arboreal in its habits; in fact that he is descended from the old-world monkeys, that he must

further, namely, to the belief that all animals and plants have descended from some one prototype." "Origin of Species," P. 484.

be classed with the quadrumana, the most immediate ancestor from which this descent can be traced, being an anthropomorphous Ape!*

Mr. Darwin having traced our descent from the Ape, carries us back for a countless number of ages, through Marsupials, Reptiles, Birds, Fishes, till he at last arrives at our most ancient progenitors, which he says resemble the young of Ascidians—the Ascidians being scarcely animals at all; they have recently been classed by some naturalists amongst the Vermes or Worms; their larvæ or young somewhat resemble

---

* "The early progenitors of man were no doubt once covered with hair, both sexes having beards; their ears were pointed and capable of movement; and their bodies were provided with a tail having the proper muscles. The males were provided with great canine teeth, which served them as formidable weapons." "Descent of Man," Vol. I., Pp. 206, 207.

tadpoles in shape, and have the power of swimming freely about.

The following may be considered as a logical description of Mr. Darwin's Genealogical Tree:—

At the bottom is Man, who may be described as an animal belonging to the Class Mammalia, of the Order of the Primates, of which he is placed at the head, in the family of the Hominidæ or Bimana. He forms the only genus of the family, and there is but one species of this genus —Homo Sapiens, " the beauty of the world, the paragon of animals."

The third stage of descent, or rather of ascent, is the Man-like Ape, represented by the Simia Satyrus or Orang Outang, a hairy animal, that is unable to hold itself upright except by clutching to the branches

Fig. 1.—The Man-like Ape.

of a tree. It inhabits the low swampy districts of Borneo and Sumatra, where it lives exclusively on fruits. When full-grown, it attains a height of about four feet four inches; its legs are very short, whilst its arms, on the contrary, are exceedingly long, reaching down to the ankles. Its facial angle is only 30°, and like all the true Apes, it has no tail.*

* I need not say that the Gorilla, the Chimpanzee, of the Gibbon, would equally well represent the Man-like Ape, as they, together with the Orang, are spoken of as the "anthropoid or latisternal apes." I have, however, selected the Orang, because, although diverging more from man, as regards the skeleton, than does any other anthropoid ape, he appears nearest to man as far as the brain is concerned, in which most important organ the Orang is man's nearest ally, as shown by the relative height of the cerebrum, the large proportion of its frontal lobe, and the high and rounded form of the skull.

Those who may desire further information as to the points of resemblance and points of difference between the anthropoid apes and ourselves, I would refer to

But between the first and third branches of this tree, that is, between the Man-like Ape and ourselves, there is a *"missing link"* —an inferred organism, for Mr. Darwin assumes that some hundreds of thousands of years ago, there was an Ape-like Man. Yes, Mr. Darwin, looking back through the dim vistas of untold ages, traverses the corridors of time, and plunging into a bygone eternity, from the dark recesses and chasms of which, lighted up only by

Mr. St. George Mivart's interesting work entitled "Man and Apes," in which the structural peculiarities of the Anthropoidea are described with great minuteness. I may, however, observe that the apes are divided into two families; Simiadæ, or apes of the Old World, and Cebidæ, which are exclusively confined to tropical America. The Simiadæ are again sub-divided into three smaller groups or sub-families :—1° Simiinæ ; 2° Semnopithecinæ ; 3° Cynopithecinæ. The first of these sub-divisions contains the Gorilla, the Chimpanzee, the Orang, and the Gibbon ; these creatures being the apes which, on the whole, are most like man.

the beams from his own distorted imagination, he drags into existence this monster of his own creation—this Ape-like Man! Professor Haeckel, of Jena, in a work entitled "The Natural History of Creation,"* has entered into very minute particulars in reference to this hypothetical being—our direct ancestor—this Homo primigenius, who, having sprung

* Natürliche Schöpfungsgeschichte, von Ernst Haeckel, Fünfte Auflage, Berlin, 1874.

Dr. Haeckel, Professor of Zoology, in the University of Jena, is justly regarded as the most eminent living representative of the doctrine of Evolution in Germany. He is a most enthusiastic admirer and devoted disciple of Mr. Darwin, whose theory he considers as "one of the greatest conquests of the human mind, and worthy to rank with the Newtonian theory of gravitation." Professor Haeckel's remarkable work has already reached a fifth edition in Germany, but as I have reason to believe that the majority of English readers are unacquainted with the peculiar views therein set forth, I shall deem it desirable to make a frequent allusion to this elaborate treatise in the following pages.

by evolution from the Anthropoid Apes, lived in the pliocene period of the tertiary age, his birthplace being a continent in Southern Asia, called Lemuria, long since submerged by the Indian Ocean! This hallowed spot he speaks of as the "so-called Paradise, the cradle of the human race" (das sogenannte Paradies, die Wiege des Menschengeschlechts.)

Of this primitive man, (Der Urmensch) he says, we as yet possess no fossil remains, but there is such an analogy between the lowest woolly-haired men (Ulotriches), and the highest Anthropoids (Menschen-affen) that it requires no great effort of the imagination to describe an intermediate type, an approximate portrait of the conjectural primitive Ape-like Man. (muthmasslich Urmensch oder Affen-mensch).

" This primitive man was very dolicephalic and prognathous; he had woolly hair, and a black or brown skin, and his body was more abundantly covered with hair than in any existing race; his arms were longer and more robust, whilst his legs were shorter and more slender than the corresponding limbs of his immediate descendant, the Homo Sapiens of the present day. When standing, his position was only semi-vertical, with the knees much bent; and he was *without articulate language.*"

"We are therefore justified," says Haeckel, "in admitting into the human pedigree, as representing the twenty-first link, the Speechless Man, (Alalus,) or the Ape-like Man, (Pithecanthropus,) a being endowed with all the essential characteristics of man, except articulate language."*

Now, it is important to remember that this assumed connecting link between Man and the Ape, is the very key-stone of the Darwinian structure. There is, however, no evidence of the existence, nor have any fossil remains ever been discovered, of this Ape-like Man; the petrified relics of extinct animals that have lived in by-gone ages have been examined, but these "*material archives of the creation*" have been searched in vain; there is no voice in the stone book of the past, not one single footprint on the sands of time, that can justify Man in his pride and presumption in attempting to bridge over the impassable gulf which separates the howling

---

* Natürliche Schöpfungsgeschichte, Pp. 597, 620.

monkey from. the being who we are told was formed in the image of his God.*

* Mr. Darwin tackles this difficulty in the following specious terms. ("Descent of Man," Vol. I., P. 200.) ": The great break in the organic chain between Man and his nearest allies, which cannot be bridged over by any extinct or living species, has often been advanced as a grave objection to the belief that man is descended from some lower form; but this objection will not appear of much weight to those who, convinced by general reasons, believe in the general principle of evolution." It will be observed that by this line of argument, Mr. Darwin takes for granted, the theory to be proved.

Further on, at P. 201, Mr. Darwin says:—" With respect to the absence of fossil remains, serving to connect man with his ape-like progenitors, no one will lay much stress on this fact, who will read Sir C. Lyell's discussion, in which he shows, that in all vertebrate classes, the discovery of fossil remains has been an extremely slow and fortuitous process. Nor should it be forgotten that those regions which are the most likely to afford remains connecting man with some extinct ape-like creature, have not as yet been searched by geologists."

Archdeacon Pratt, animadverting on the above passage, remarks:—" If we knew that the theory is

And, forsooth, it is upon evidence like this, that we are asked to forego the cherished traditions of our forefathers, and to substitute the audacious theories of yesterday for a record of creation which, for more than thirty long centuries, has successfully resisted the battering-ram of infidelity and unbelief, and for three thousand years, has braved the battle and the breeze of scepticism and doubt.

Let us now continue the study of our Genealogical Tree. From the man-like Ape, we are carried up to the Catarrhine or Old World Monkeys, a good specimen of which is seen in the Macacus Radiatus, or Bonnet Monkey, a member of the

---

true, we should be sanguine that, some day, proof would be found in fossils; but as the whole is a gratuitous hypothesis, the entire absence of fossil proof is a stern rebuke to the speculators."

Fig. II.—Catarrhine or Old World Monkey.
(Macacus.)

Figures I, II, VII, and VIII, are reproduced with permission from Mr. St. George Mivart's work, "Man and Apes."

third sub-family or Cynopithecinæ, a creature well known in this country, being frequently brought over by soldiers and sailors. It is less gentle and docile than some other monkeys, being a snappish, irritable animal, and when not indulged, is given to mischievous and spiteful tricks; it is provided with a tail.

Mr. Darwin next traces us to the Macropus Major, or Kangaroo, one of the Marsupials, and from this dignified beast, he carries us through reptiles and other organisms to the fishes, which we may suppose to be represented by the Sturgeon, Acipenser Sturio, when our ancestors swam in proud majesty in the azure waters of the sea. From the Sturgeon, we are conducted to the Amphioxus or Lancelet, the lowest known vertebrate animal, a creature looking very much like a piece of jelly. This little animal is remarkable for its negative properties, having neither brain, head, nor heart; it has been described by a modern anatomist as a "headless, heartless fish, without red blood."* Professor Haeckel evidently regards the Amphioxus as representing one of the most important stages in man's pedigree, remarking that "the study of this interesting little animal throws great light upon the roots of our genealogical tree, forming as it does the line of demarcation between the vertebrates and the invertebrates." He calls it the last of the Mohicans, (der letzte Mohicaner) and

* "The possession of a heart and of red blood is common to all vertebrates as well as to man, with one solitary exception, the Amphioxus or Lancelet alone having colourless blood and a simple cylindrical vessel in place of a heart." "Mivart, Lessons in Elementary Anatomy," P. 12.

says that " the study of its comparative anatomy and ontogenesis proves to a certainty the former existence of animals without a skull and without a brain amongst the ancestors of Man!"*

By the next and last step of the Darwinian ladder, we are carried up to the Ascidian, which is described as an invertebrate hermaphrodite marine creature, permanently attached to a support, and immovably fixed at the bottom of the sea by root-like appendages, whereas its near relative, the Amphioxus, can swim freely like a fish. It belongs to the Molluscoida of Huxley, a lower division of the great kingdom of Mollusca. The Ascidian (ἀσκός, a skin bottle) consists of a simple tough leathery sac, with two small projecting orifices, and its appearance very

* Haeckel op cit, Pp. 508, 584.

much resembles a double-necked jar. "At first sight," says Professor Huxley, "you might hardly suspect the animal nature of one of these singular organisms, when freshly taken from the sea; but if you touch it, the stream of water which it

Fig. III.—The Ascidian.
Our Pre-Historic Ancestor.

squirts out of each aperture reveals the existence of a great contractile power within." Of the two apertures, A serves as a mouth, and is often surrounded by a circle of tentacles; B is the anal orifice, and C is the base of attachment, by which the animal fastens itself to a bit of seaweed or to a rock. This interesting creature is here represented, in order to enable one to form some idea of man's very remote ancestors.* The engraving is taken from Professor Huxley's "Elements of Comparative Anatomy," the author having kindly permitted me to copy it. Thus the lofty faculties of Man were once in embryo in a thing like a tadpole! The mind of Newton once lay hid in a creature which "hardly appeared like an

* I have not thought it desirable minutely to describe the long line of diversified forms through which Mr. Darwin ultimately traces us up to our common ancestor, the Ascidian; for a more detailed description of Man's Genealogy from the Darwinian point of view, I would refer the reader to an interesting and highly scientific treatise, by Dr. Bree, of Colchester, entitled "Fallacies of Darwinism," from which I have obtained most valuable information in the compilation of this work.

animal—which consisted merely of a simple tough leathery sac, and which stuck to a bit of sea-weed that it might not be carried away by the tide." *

Thus far Mr. Darwin, but my description of the object, aim, and end of the Evolution theory, as applied to the descent

* Dr. Payne Smith, Dean of Canterbury, has the following reflections upon the Ascidian descent of man. " What an alarming thought, that at a period separated from us by such vast geologic ages, that, according to the nebular hypothesis, held by so many of our leading astronomers as a probable theory, this whole universe was a mass of heated vapour ; what an alarming thought that the very existence of man should have depended upon a jelly bag sticking to a stone and sucking up water ! Alas ! there was then no water, no stones, no jelly bags, and therefore there are now no men ! Man escapes, poor thing, from his humble parentage : he need not feel his ears to find the proof of his monkey-hood : but his escape costs him dear. What with astronomy and biology, men of science between them have cleared us out of existence. Scientifically, man is no more." "Modern Scepticism;" P. 150.

of man, would be incomplete without a further reference to Professor Haeckel's views. Mr. Darwin, as we have seen, is content with tracing man's descent from an Ascidian Mollusk, and he is also satisfied with deriving all animals and plants from about eight or ten progenitors, whereas, his most valiant disciple, Professor Haeckel, goes much further back, through a complete family tree of twenty-two branches, and having reached Mr. Darwin's Ascidian, he carries us seven stages higher up, through sponges, diatoms, worms, and other organisms, till he eventually traces us all to one *primordial germ*—a Moner, produced by self generation (Archigony) from inorganic matter during the Laurentian period.

This Moner—μονήρης—the lowest imaginable grade of organic individuality, he describes as a formless, structureless, slimy atom, (Schleimklümpchen) composed of an albuminoid carbonaceous matter, as homogeneous as an inorganic crystal. Although when in a state of repose, it only consists of a little ball of slime or mucus, either invisible to the naked eye, or if visible, only of the size of a pin's head, still it is endowed with the two fundamental organic functions of nutrition and reproduction.

" These first ancestors of man," says Haeckel, " were as simple as possible. They were organisms without organs, like our present monera, consisting merely of little shapeless lumps of a slimy albuminous material (protoplasm). These organisms never attained to the form of a cell, but were always mere ' *cytodes*,' being devoid of any nucleus. The first of these monera sprang by spontaneous generation,

(Urzeugung) at the commencement of the Laurentian period, from inorganic compounds—simple combinations of Carbon, Carbonic Acid, Hydrogen, and Nitrogen."*

The Monera are further described as being neither plants nor animals, but belonging to a third primary division of the living world, to which Haeckel has given the name of Protista.

As the history of the Protistic kingdom may be a novelty to many of my readers, I shall not deem it irrelevant to my subject to enter into some details in reference to it. The Protista form an organic group which cannot naturally be classed either in the animal or vegetable kingdom; there being in their exterior form, in their intimate structure, and in their vital phenomena, such a singular mixture of

* Natürliche Schöpfungsgeschichte, P. 578.

animal and vegetable properties, that they have been respectively claimed both by the botanist and by the zoologist.

The Primordial organisms which constitute the Protistic kingdom are divided into the following eight groups:—

1°, The Monera. 2°, The Amœboida or Protoplasta. 3°, The Flagellata. 4°, The Catallacta. 5°, The Labyrinthulæ. 6°, The Diatomaceæ. 7°, The Myxomycetes or mucus-fungi. 8°, The Rhizopoda.

The accompanying illustration (Fig. IV) represents the most interesting member of this Protistic kingdom, The Moner, "the first ancestor of Man," and also shows the mode of reproduction observable in these elementary organisms, which is by segmentation; that is, when one of these little corpuscles has acquired a certain size by the absorption of albuminoid matter, it begins to show a tendency to divide into two parts; a central constriction occurs, resulting eventually in a separation into two halves, each half becoming henceforth a distinct individual, possessed of all the properties of the parent Moner.

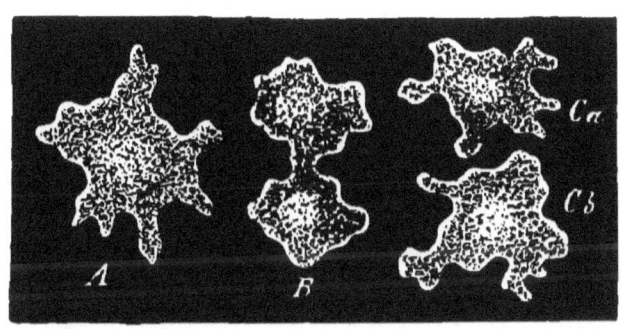

Fig. IV.—The Moner.
Man's First Ancestor.

A. Is the entire Moner. B. The same corpuscle divided into two halves by a median furrow. Ca. Cb. The two halves have become separated from each other, and now constitute distinct and independent individuals, manifesting the same vital phenomena of nutrition and reproduction as the organism of which they originally formed a part.

"In certain instances," says Haeckel, "the Monera sub-divide into more than two parts, and in some species they separate into a great number of mucous globules, which by simple growth acquire the volume of their parents." He then goes on to say that "this most simple mode of reproduction, by scissiparity or self-division, is the same by which *cells* are re-produced—those rudimentary organic units, by the agglomeration of which almost all organisms are constituted, not excepting even the human body. Each organic individual is always composed of a great number of cells, and each cell is, to a certain extent, an individual organism —a being of primal order."

Fig. V represents the second group of the Protistic kingdom—the Amœboida or Protoplasta. Here it will be observed we have advanced a step, and have attained to the dignity of a true *cell*, the Amœba being a monocellular organism containing a *nucleus*. The mode of reproduction differs from that which obtains in the Monera, for although occurring by scissiparity or self-division, it is the *nucleus* itself which separates into two halves; the cell substance eventually divides, thus forming two new cells resembling the mother-cell. Under certain conditions of rest, etc., the Amœba assumes a globular form, and becomes invested with a cell-membrane or cyst, as is shown in the accompanying engraving.*

* The engravings Nos. 4, 5, and 6 are taken from Haeckel's original work, the Professor having most

A is the Amœba, a simple spherical cell, consisting of protoplasm (c), containing a nucleus (b), and a nucleolus (a), the whole organism being enclosed in a cell-membrane (d). B. The Amœba has ruptured and escaped from the cyst. C. The nucleus has separated into two nuclei, and the Amœba itself is constricted by a median furrow. D. The division is complete, and two independent cells are formed, each with its proper nucleus. (Da Db). The white corpuscles of the blood of man and of animals, says Haeckel, cannot be distinguished from these Amœbæ.

The study of this elementary form has evidently great attraction for the German Professor who says that, after the Monera, the Amœbæ are the most important of all

courteously permitted me to reproduce them from blocks supplied to me by his publisher at Berlin.

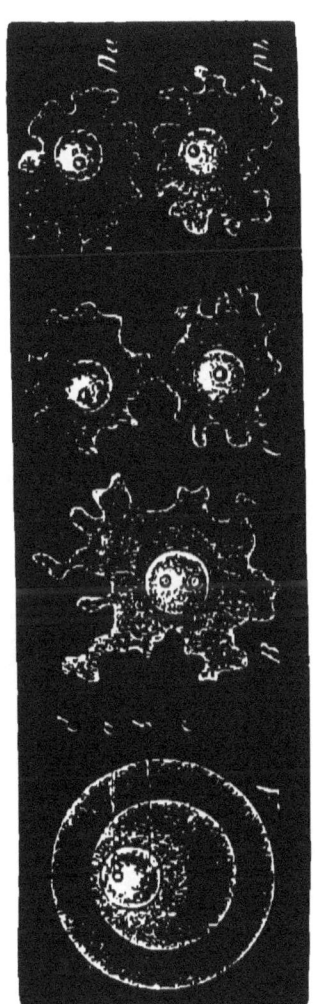

Fig. V.—The second group of the Protistic Kingdom.

The above engraving shows the mode of reproduction in a monocellular organism, by segmentation or self-division, as exemplified in the Amœboida or Protoplasta.

organisms in a biological, and especially in a genealogical, point of view.

The third group of this kingdom of Primitive Forms is that of the Flagellata, which are organisms consisting of simple cells, living in fresh or in salt water.

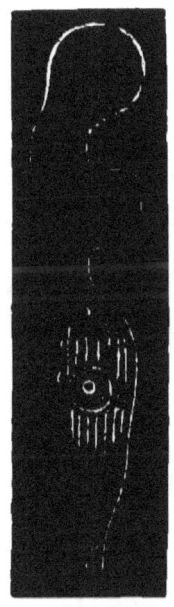

Fig. VI.—A Flagellate.

They are characterised by a flagelliform appendage, such as is represented in figure VI., which serves for the purpose of rapid movement in the water; they possess also a nucleus and a nucleolus, as is shown in the engraving. It is to these organisms that is due, in a great measure, the phosphorescence of the sea, and their presence in large numbers imparts a green colour to our ponds.

I do not propose giving details of the five other groups of the Protistic Kingdom, the description of which forms one of the most interesting features of Professor Haeckel's elaborate treatise. The greater part of the Protista, he says, live in the sea, some swimming on the surface, others crawling at the bottom, or permanently attaching themselves to rocks, shells, or plants. They are so small that, for the most part, they can only be discerned by the aid of the microscope. I must not omit the startling assertion that "all the

Protista have a soul (eine Seele) as well as all animals and all plants!"*

In tracing man's pedigree, Haeckel divides the earth's history into five great periods. The first is the Archolithic or Primordial Age, during the early part of which, our first ancestor, the Moner, appeared. After passing through various successive elementary forms, at the eighth stage of descent, he reaches the Ascidian, at which point, the ancestors of man pass from the invertebrate to the vertebrate type;

---

* Eine Seele besitzen alle Protisten, so gut wie alle Thiere und wie alle Pflanzen. Die Seelenthätigkeit der Protisten, äuszert sich in ihrer Reizbarkeit, d. h. in den Bewegungen und anderen Veränderungen, welche in Folge von mechanischen, elektrischen, chemischen Reizen u. s. w. in ihrem contractilen Protoplasma eintreten. Wie bei allen übrigen Organismen, so sind auch bei den Protisten die Seelenthätigkeiten auf Molekular-Bewegungen im Protoplasma zurückzuführen. Natürliche Schöpfungsgeschichte, P. 393.

D

three more evolutionary changes, however, are observed during this first period, and at the eleventh stage, the Moner has become developed into a Selacian or shark-like fish. Then begins the second or Palæolithic Age, during which, on reaching the fourteenth stage, he has attained the dignity of a Triton. In the third or Mesolithic Age, he moves three steps higher up, and becomes a Kangaroo. In the course of the fourth or Cainolithic Age, four more stages are passed through, and he successively assumes the form of the Lemur, the Old World Monkey, the Anthropoid Ape, and the Speechless or Ape-like Man. In the fifth or Quaternary Age, or period of human civilisation, we arrive at the twenty-second stage of evolution, represented by Man endowed with the faculty of Articulate Language.

# CHAPTER II.

"Non enim est e saxo sculptus aut e robore
dolatus, habet corpus, habet animum, movetur
mente, movetur sensibus."
CICERO. *Acad. Prior.* ii. 36.

*Evolution theory contrasted with the Scriptural account of the origin of Man—The Monistic and Dualistic hypotheses—The Primordial Germ—What brought it into existence?—Archdeacon Pratt's and Dr. Reichel's views—Mr. Disraeli on Evolution.*

In the preceding chapter, I have endeavoured to epitomise the views of Mr. Darwin and of Professor Haeckel as to the origin of the human race; the former tracing man's pedigree up to the Ascidian Mollusk, whilst, according to the German naturalist, the first ancestor of man was a much more simple organism—

a shapeless, structureless, slimy Atom, produced by spontaneous generation.

The theory of Evolution abolishes the idea of creation, in the ordinary sense of the term. It, at most, concedes to Nature the faculty of causing one species to spring from another, and it consequently excludes all direct, personal, and miraculous intervention of a creating power. Instead of assigning existing species to the creative act of the Maker of all things, the Evolutionist imagines them to be derived by natural causes out of previous forms, and these again out of others, up to an original germ or protoplasm. Evolution, in fact, means that a system of laws and forces has been set in motion which produces certain results without any interference or assistance from a superintending power. It assumes that advances in development have taken place not by design, but by accident, or the force of circumstances. Its fundamental proposition, according to Huxley, is, that "the whole world, living and not living, is the result of the mutual interaction, according to laws, of the forces possessed by the molecules of which the primitive nebulosity of the universe was composed."

It may be well to contrast the Evolution theory, with the Scriptural account of the origin of Man, which may be thus summarised:—

1°—The universe with all it contains owes its existence to the will and power of God; matter is not eternal, nor is life self-originating. The Deity has endowed matter with properties and forces, which He upholds, and in accordance with which

He works in all the ordinary operations of His providence.

2°—Man's body was formed by the immediate intervention of God. It did not grow; nor was it produced by any process of development.

3°—The soul was derived from God. He breathed into Man "the breath of Life,"* that is that life which constituted

---

* Whilst this chapter is passing through the press, a learned Hebrew scholar, and one of the great theological writers of the day, has reminded me that in the original of Genesis ii. 7, the words are *nishmath hayim*, breath of *lives*, not of *life*. "Most ancient commentators," he says, "notice the force of the plural, as intimating that not only the animal, but the intellectual and moral life of man were conveyed by that Divine insufflation; and Josephus himself (rather an interpreter of a rationalistic caste of thought) says, that both *soul* and *spirit* were breathed into the body of man. If the animal life of man could possibly be conceived to have been developed from the life of lower species, whence did his intellectual and still more his moral life come?"

him a man—a living being bearing the image of God.

4°—Of the various creatures summoned into existence prior to the creation of Man, each is said to be "*after his kind,*" words which seem to imply that, from the first, each species was distinct from the other. "It was a *kind*" by itself.

The above view of creation has been called the *dualistic hypothesis*, according to which, organic matter is considered to be the premeditated work of a Creator, acting in accordance with a fixed plan, and man, at the very first moment of his existence, was separated from the highest brute by as impassable a gulf as that which exists between them at the present time.

The theory of Evolution, which has also been termed the *monistic hypothesis*, attributes all vital phenomena to mechanical causes, either physical or chemical; all animal and vegetable species of the present and of past ages are merely "the posterity slowly modified and transformed of one or more very simple original ancestral forms, issued by spontaneous generation from inorganic matter." *

A primordial germ, with no inherent intelligence, and by the slow operation of unintelligent physical causes, develops, under purely natural influences, into all the infinite variety of vegetable and animal organisms, with all their complicated relations to each other, and to the world around them. All living things,

* Natürliche Schöpfungsgeschichte, P. 106. Haeckel calls the monistic hypothesis mechanical or causal, as distinguished from the dualistic, which he calls teleological or vital, according to which, each animal and vegetable species is the product of an "*incarnate creative thought.*" (verkörperten Schöpfungsgedanken.)

from the lowly violet to the giant redwoods of California, from the microscopic animalcule to the Mastodon and the Dinotherium, one and all have sprung from this same primordial germ.* The

* A recent theological writer inquires :—" What brought this primordial form into existence? The pushing back of its first appearance further and further into past time, ages before ages, and ages before them, does not get rid of the question, How came this form into existence? A form, too, possessing such marvellous properties, as to give birth to all the varieties of organisation which the vegetable and animal kingdoms exhibit. God must have created it. If, then, the Almighty created one such form, why could He not have created several? What necessity is there in the nature of things for tracing up the genealogy of all organic beings to one form only?" "Scripture and Science not at Variance," by J. H. Pratt, Archdeacon of Calcutta, P. 228.

Dr. Reichel, in further development of this subject, after ridiculing the idea of the Hindoo who says that the world rests on an elephant, and the elephant upon a tortoise, and then thinks that he has given a sufficient account of things without telling us what the tortoise rests on,

Evolutionist not only asserts that all this is due to natural causes, without purpose or without design, but he argues against the intervention of mind anywhere in the process. God, says Lamarck, created matter; God, says the Evolutionist, created the unintelligent living cell ; both say that after the first step, all else follows by natural law, without purpose and without design. From certain primary elements, such as

goes on to say :—" make the chain of finite causes as long as you like; multiply its links (each link a Universe) as often as you please ; this chain must have an end ; and by the very necessities of thought you are driven to acknowledge that at its end there must be one ultimate cause, different from all other causes, existing by the necessity of its own nature before all other causes, and which, because it exists by inherent necessity, can never cease to exist. Thus arrangement, plan, design, are only pushed a little further back by the evolution theory : they are not got rid of."
" Norwich Cathedral Argumentative Discourses," *Series* vi., Pp. 8, 12.

soil, and stream, and wind, the solar heat, it seems, would be sufficient to undulate matter into Man, and out of such airy materials as Oxygen and Hydrogen, Carbon and Nitrogen, with a little Phosphorus and Lime thrown in, the Evolutionist would erect society, with its science and its government, its art and its religion!

"Many who hold the evolution hypothesis," says Tyndall, "would probably assent to the position, that at the present moment, all our philosophy, all our poetry, all our science, all our art—Plato, Shakespeare, Newton, and Raphael—are potential in the fires of the sun." God, it would seem, hundreds, or perhaps thousands, of millions of years ago, called this primordial germ into existence, and since that time, has had no more to do with the universe than if *He* did not exist. According to this theory, the Supreme Being would be regarded in the light of a skilful mechanician, who, after constructing the universe, and setting it at work, withdrew himself thenceforth from all interference with it, as completely as a clockmaker does, in the instance of a clock which he exports to a foreign country, or as a ship-builder after the ship is constructed, launched, and is far away on the wide ocean. The Deity having created matter, and endowed it with certain properties, does nothing more, retires into inactivity, and without any control or interference on his part, hands over this product of his creative power to the guidance of physical laws. "Human nature exists potentially in mere inorganic matter, and a chain of spontaneous derivation connects incandescent molecules or star-dust with the world and man himself." Everything in creation has arisen from a fortuitous concourse of atoms, and life itself is said to be the product of a certain disposition of material molecules, the matter of life being composed of ordinary matter, and differing from it only in the manner in which its atoms are aggregated. Thus, life, mind, and all the infinite diversities and marvellous organisms of plants and animals, from the lowest to the highest, are due to the operation of unintelligent physical causes. The earth is assumed to be pregnant with the germs of all living organisms, which are quickened into life under favourable circumstances; in the bosom of inorganic nature are various dormant forces, which, at certain times and under certain conditions, spring into action and develop into a plant or an animal, just as we see a crystal formed by virtue of certain chemical affinities; and at a particular conjuncture in the world's history, and from the coincidence of certain special conditions, Man appeared as the final product of the operation of nature's laws.*

Our great statesman-novelist, Mr. Disraeli, in the conversation between Lothair and the Syrian, as they sat gazing on the wondrous scene afforded by the morning view of Jerusalem from the

* In the above summary, I have endeavoured impartially to represent the views of the different leaders of modern thought. I am quite aware that my description applies rather to the Darwinian school than to its founder, whose doctrine, as originally promulgated, merely implies a belief in the origin of species by Natural Selection, but as the words Evolution and Darwinism are now used interchangeably, it is convenient to adopt the latter as a generic term.

Mount of Olives, has beautifully demonstrated the absurdity of the above position, remarking that nothing can be more monstrous than to represent a Creator as unconscious of creating. " There must be design," says the Syrian, " or all we see would be without sense, and I do not believe in the unmeaning. As for the natural forces to which all creation is now attributed, we know they are unconscious, while consciousness is as inevitable a portion of our existence as the eye or the hand. The conscious cannot be derived from the unconscious." Lothair having expressed a wish that he could assure himself of the personality of the Creator, but that he had been told that such an idea was unphilosophical, the Syrian thus replies:—"Is it more unphilosophical to believe in a personal God, omnipotent and omniscient, than in natural forces unconscious and irresistible? Is it unphilosophical to combine power with intelligence? Goethe, a Spinozist who did not believe in Spinoza, said that he could bring his mind to the conception that, in the centre of space, we might meet with a monad of pure intelligence. What may be the centre of space I leave to the dædal imagination of the author of 'Faust;' but a monad of pure intelligence, is that more philosophical than the truth, first revealed to man amid these everlasting hills—*that God made man in His own image?*" \*

---

\* Lothair, Vol. III., Pp. 179, 183.

# CHAPTER III.

"Ceux qui ont dit qu'une fatalité aveugle a produit tous les effets que nous voyons dans le monde ont dit une grande absurdité ; car quelle plus grande absurdité qu'une fatalité aveugle qui aurait produit des êtres intelligents."

MONTESQUIEU, *De L'Esprit des Lois.*

*Sentimental opposition deprecated—Broca, Max Müller—No evidence of transmutation of species within the historic period — Flourens—Animal Kingdom of Aristotle, the same as that of our day. Plea of the Imperfection of the Geological Record considered—Haeckel, Duke of Argyll, and Mivart— Professor Agassiz on the Immaterial Principle.*

In considering the validity of the arguments which can be adduced for or against the theory of Evolution, I desire to approach the subject in a spirit of toleration and impartiality, and I trust I shall say nothing in this essay to justify my being classed amongst those whom Mr. Darwin describes as "curiously illustrating the blindness of pre-conceived opinion," or amongst those whom Professor Huxley represents as "contenting themselves with smothering the investigating spirit under the feather-bed of respected and respectable tradition." I deprecate all idea of stirring up the *odium theologicum*, being fully conscious of the futility of attempting to check an unwelcome and distasteful theory by means of ecclesiastical censures. I consider the doctrine of Evolution as a legitimate subject for scientific inquiry; I recognise the deep knowledge of natural history which the "Descent of Man" displays; I fully endorse the terms of high commendation in which its literary merit has been acknowledged, and from its charm of style and elegance of diction, I am not surprised that it has become equally popular in the drawing-room of the votary of fashion, as in the study of the naturalist and the theologian.

I should not reject the Darwinian view of the origin of man, from any fancied notion that its adoption was derogatory to our dignity, and inconsistent with Man's position in the order of Nature, a notion which was evidently held by the poor deluded creature whose suicide was lately recorded in the public papers, and upon whose person was found a document, stating that his existence was no longer to be tolerated, since Mr. Darwin's discovery that he was descended from a monkey. Instead of sympathizing with the views of this unhappy victim of prejudice and folly, I fully echo the sentiment of the naturalist who said that he would prefer being descended from a good honest monkey, to being obliged to avow himself the offspring of certain fanatical enemies of scientific knowledge and progress.*

---

*Professor Broca, of Paris, has developed the above idea in the following terse and eloquent language, the force of which I will not impair by a translation:—"Je ne suis pas de ceux qui méprisent les parvenus. Je trouve plus de gloire à monter qu' à descendre, et si j'admettais l'intervention des impressions sentimentales dans les sciences, je dirais, comme M. Clarapède, que j'aimerais mieux être un singe perfectionné qu' un Adam dégénéré. Oui, s'il m'était démontré que mes humbles ancêtres furent des animaux inclinés vers la terre, des herbivores arboricoles, frères ou cousins de ceux qui furent les ancêtres des singes, loin de rougir pour mon espèce de cette généalogie et de cette parenté, je serais fier de l'évolution qu 'elle a accomplie, de l'ascension continue qui l' a conduite au premier rang, des triomphes successifs qui l'ont rendue si supérieure à toutes les autres. Je me réjouirais en songeant que mes descendants, poursuivant indéfiniment l'œuvre splendide du progrès, pourraient s'élever au-dessus de moi autant que je m'élève au dessus des singes, et réaliser enfin cette promesse du serpent de la Genèse: *Eritis sicut deos !*" " Sur le Transformisme," P. 2.

After all, the question is not whether the theory of the Simian descent of man is palatable, or in accordance with our conventional notions, but simply and solely whether it is true. " Appeals to the pride or humility of man" says Professor Max Müller, " to scientific courage or religious piety, are all equally out of place. If it could be proved that our bodily habitat had not been created in all its perfection at the first, but had been allowed to develop for ages before it became fit to hold a human soul, should we have any right to complain? Do we complain of the injustice of our having individually to be born or to die; of our passing through the different stages of embryonic life; our being made of dust, that is, of exactly the same chemical materials from which the bodies of animals are built up? Fact against fact, argument against argument, that is the rule of scientific warfare, a warfare in which to confess oneself convinced or vanquished by truth is often far more honourable than victory." *

Whatever, moreover, may have been the remote origin of man, we can cheer ourselves with the thought, that for ages he has possessed a history of his own; he has filled the world with monuments of his ambition, skill, and genius; and he is the sole actor in a drama where other animal beings play only an accessory part.

In my description of Man's Genealogical Tree, I had occasion to speak of the "Missing Link," or the absence of the intermediate forms between man and his

---

* Lectures on Mr. Darwin's Philosophy of Language. "Frazer's Magazine," June, 1873, P. 665.

supposed progenitors, either in a living state or in a fossil condition. In furtlter development of this subject, I would observe, that, in the earliest description we have of man, we find him separated from the highest brute by as wide a gulf as that which now exists between them; the oldest human skulls are not materially inferior in capacity to those of man at the present day, as may be seen by a visit to the Anthropological department of our museums; and Professor Huxley in describing the Engis skull, which according to Sir Charles Lyell belonged to a contemporary of the Mammoth, says, that "It is a fair average skull, which might have belonged to a philosopher, or might have contained the thoughtless brains of a savage." *

* " Man's Place in Nature," P. 156.

The embalmed records of three thousand years, the figures of animals and birds engraved upon the ancient Egyptian tombs and obelisks, "those hoary monuments of early science," show that there has been no beginning of a transition of species during the long space of thirty centuries. During the whole of the historical period, species have remained unchanged, they are precisely what they were thousands of years ago; there is not the slightest indication of one passing into another, or of a lower advancing to a higher; moreover, each species has manifested in its capabilities, as well as in its organisation, certain indelible peculiarities, which have been transmitted from age to age. There is an entire and acknowledged absence of all evidence of transmutation, and none of the transition points or links of connection between one species and another are anywhere discoverable, thus verifying the aphorism of M. Flourens. "*Les espèces ne s'altèrent point ; ne passent point de l'une à l'autre ; les espèces sont fixes.*" In justification of the above statement, M. Flourens says, "It is two thousand years since Aristotle lived ; guided by comparative anatomy, Aristotle divided the animal kingdom as Cuvier has done in our own day. There were in it viviparous quadrupeds or mammals, birds, oviparous quadrupeds or reptiles; there were also fish, insects, crustacea, mollusks, radiates, or zoophytes. The animal kingdom of Aristotle is the animal kingdom of to-day. The animals which Aristotle has described, are recognized in the present time, even to the minutest particular."\*

\* Examen du livre de M. Darwin sur l'origine des espèces par P. Flourens, Membre de l'Académie Française, P. 22.

The only answer to the difficulty thus presented is, that the change of species is so slow a process, that no indications can be reasonably expected in the few thousands of years within the limits of history. When it is objected that geology presents the same difficulty, and that the genera and species of fossil animals are just as distinct as those now living, we are told that the records of Geology are too imperfect to give us full knowledge on this subject, and that innumerable intermediate and transitional forms *may* have passed away, leaving no trace of their existence; or, forsooth, the fossil remains of traditional links may still be entombed in some undisturbed portion of the crust of the earth, indeed, Mr. Darwin lays great stress on the fact that those regions which are the most likely to afford remains connecting man with some extinct ape-like creature, have not as yet been searched by geologists. Professor Haeckel dilates at considerable length upon this imperfection of the Geological Record, but whilst admitting that the "archives of creation," (Schöpfungsurkunde) are most incomplete, he endeavours to explain that the palæontological gaps are due to the fact that but a small portion, perhaps not a thousandth part, of the surface of the globe has been geologically explored. He reminds us that three-fifths of the surface of the globe is submerged, and that consequently we can never know what fossils of primitive ages may be buried at the bottom of the sea, although possibly they may be studied many thousand years hence, when, by reason of gradual changes, the bottoms of the present seas shall have become dry ground.* If we say that the Ape, during the historical period, extending over thousands of years, has not made the slightest approximation towards becoming a man, we are told, Ah! but you do not know what he will be in ten millions of years;

* Natürliche Schöpfungsgeschichte, Pp. 355, 356.

The Duke of Argyll takes a much more logical and practical view of this subject:—" It is true," says he, "that the geological record is imperfect, but as Sir Roderick Murchison has long ago proved, there are parts of that record which are singularly complete, and in those parts we have the proofs of Creation without any indication of Development. The Silurian rocks, as regards Oceanic life, are perfect and abundant in the forms they have preserved, yet there are no Fish. The Devonian Age followed, tranquilly and without break; and in the Devonian Sea, suddenly Fish appear—appear in shoals, and in forms of the highest and most perfect type. There is no trace of links or transitio na forms between the great class of Mollusca and the great class of Fishes. There is no reason whatever to suppose that such forms, if they had existed, can have been destroyed in deposits which have preserved in wonderful perfection the minutest organisms. So much for the Past." "Primeval Man." P. 44.

to which surely, a suitable rejoinder would be, to ask, how much is ten millions time nothing ?*

There is one consideration in connection with this branch of the subject which has been urged with great force by the author of *Homo versus Darwin:*—" Why are enormous periods of time required for the production of new species, but that there may be successive generations, each of

* Mr. St. George Mivart in discussing the relation of species to time observes :—" The mass of palæontological evidence is indeed overwhelmingly against minute and gradual modification. Not only are minutely transitional forms generally absent, but they are absent in cases where we might certainly *a priori* have expected them to be present. Had such a slow mode of origin, as Darwinians contend for, operated exclusively in all cases, it is absolutely incredible that birds, bats, and pterodactyles should have left the remains they have, and yet not a single relic be preserved, in any one instance, of any of these different forms of wing in their incipient and relatively imperfect

which may be supposed to have advanced on its predecessors? Now it is clear that, in the case of numerous animals, the period of time required for this purpose would be much less than in the case of Man. We may suppose that three generations of men are produced in a century. This would give ninety generations in 3,000 years, which may be regarded as the historic period in connection with this subject. But, within the same period, we

functional condition! Thus we find a wonderful (and on Darwinian principles an all but inexplicable) absence of minutely transitional forms. All the most marked groups, bats, pterodactyles, chelonians, ichthyosauria, anoura, &c., appear at once upon the scene. Even the horse, the animal whose pedigree has been probably best preserved, affords no conclusive evidence of specific origin by infinitesimal, fortuitous variations; while some forms as the labyrinthodonts and trilobites, which seemed to exhibit gradual change, are shown by further investigation to do nothing of the sort." "Genesis of Species," Pp. 128, 129, 142.

must have had not less than 3,000 generations of those numerous species of creatures which produce a fresh progeny every year, or even oftener than that. There have thus been 3,000 successive generations of many of the lower animals within a period during which men may have been expected to observe and record any remarkable changes occurring among them. What then is the sum of the changes which Mr. Darwin is able to point to within the historic period as tending to prove his hypothesis? It amounts absolutely to nothing! Take the case of any species of animal which produces young within a year of its birth. We have reference in the writings of ancient naturalists to many of them. We have pictures of them on ancient monuments. We find skeletons of them in

ancient tombs, and in mounds and caves.
There are thus many animals living now,
which can be compared with their progenitors of the 3,000th generation back.* Can
Mr. Darwin show, then, in the case of any
one of them, that, by successive variations
accumulated during 3,000 generations, it

---

* Professor Haughton, in a lecture recently delivered at Trinity College, Dublin, appealed to his knowledge of natural history in corroboration of the above view, selecting for illustration two animals, about which he said Mr. Darwin's mind seemed to be particularly troubled—the Goose and the Cat. "The Assyrian inscriptions," he says, " show that the goose of that period was identically the same as that which we now eat for our Christmas dinner. The cat in 5,000 years has not varied in the slightest degree. Geology also is opposed to the evolution theory, for monkeys found in the fossil strata were as perfect monkeys as those now roaming the forests of Africa, the physical structure of these fossil monkeys being the same as their successors of the present day. There is, in fact, no proof that variation has ever gone on until it has resulted in the production of a new species."

has sensibly advanced towards some higher form? Can he show that 3,000 generations have in any instance, done aught towards proving the truth of his hypothesis? It appears that he cannot point to a single case as yielding him support. Three thousand generations have done literally nothing for his hypothesis. If so, neither would 30,000 nor 300,000, for if you multiply nothing by a million, it will be nothing still."*

I see nothing in the doctrine of evolution, as applied to the origin of man, that is inconsistent with *Natural* Religion. We know that in intra-uterine life, we pass through a preparatory stage which we can but imperfectly realise and understand, and therefore we can readily admit that

* "Homo versus Darwin," by W. P. Lyon, B.A., P. 138; a most thoughtful, logical, and philosophical contribution to the Anti-Darwinian literature.

the Creator, if He had chosen, could have endowed us with a previous existence in the form of a less perfect animal than man; I say, the Darwinian hypothesis of the origin of man is not inconsistent with *Natural* Religion, but it is directly opposed to *Revealed* Religion, which tells us that " God formed man of the dust of the ground, and breathed into his nostrils the breath of life, and man became a living soul." Moreover, a belief in the progressive development of man from any inferior animal whatever, is absolutely incompatible with a belief of the existence in man of an immortal spirit; for, as stated by a thoughtful writer, " by no conceivable process, can that which is essentially not material, be developed from any combination of mere material elements." *

* " Faith and Free Thought," P. 57.

My intention is not to attempt to enter into a general criticism of the validity of the arguments which can be adduced for or against the Darwinian theory; this would lead me far beyond the limits within which I propose to confine this essay; moreover, this has been done over and over again by far abler hands than mine. I propose to test Darwinism mainly, however, in reference to its bearings upon the faculty of Articulate Language. Before entering upon the subject of Language, it is desirable to make a brief review of the "Descent of Man" itself, for although I have already discussed its author's doctrine in general terms, it is important to analyse a little more closely the exact line of argument adopted in this work.

Those who have read the "Descent of

Man," will remember that the author begins by saying that he who wishes to decide whether man is the modified descendant of some pre-existing form, would probably first inquire whether man varies, however slightly, in bodily structure, and in mental faculties ; and if so, whether the variations are transmitted to his offspring in accordance with the laws which prevail with the lower animals. He then proceeds to compare the bodily structure of man and that of the lower animals, remarking that all the bones in his skeleton can be compared with the corresponding bones in the monkey, bat, or seal ; that it is the same with his muscles, nerves, blood-vessels, and viscera,—in fact, he shows that there is a remarkable correspondence between man and the higher mammals, especially the ape, in the structure of the brain and other parts of the body. He then calls attention to the fact that man is liable to receive from the lower animals, and to communicate to them, certain diseases, as hydrophobia, small-pox, the glanders, &c., a fact which he says proves the close similarity of their tissues and blood, both in minute structure and composition, far more plainly than does their comparison under the best microscope, or by the aid of the best chemical analysis. He then points out the resemblance between man and other animals in their embryonic condition, remarking that man is developed from an ovule, about the 125th of an inch in diameter, which differs in no respect from the ovules of other animals, and that the embryo itself at a very early period can hardly be distinguished from that of other members of the vertebrate kingdom.* It is, in short, says he, scarcely possible to exaggerate the close correspondence in general structure, in the minute structure of the tissues, in chemical composition and in constitution, between

* It is an established fact in natural history, that all animals may be traced to an ovule or simple little cell; but although no difference between these various cells may be discernible by our present means of investigation, the issue clearly shows that there must be an essential difference, for the ovum of a dog invariably becomes a dog; that of an ape becomes an ape; and that of a man becomes a man. Professor Hodge, in speaking of this subject, says "the germs of a fish and of a bird are indistinguishable by the microscope or by chemical analysis; yet the one, under all conditions, develops into a fish and the other into a bird. Why is this? There is no physical force, whether light, heat, electricity, or anything else, which makes the slightest approximation to accounting for this result"

Another American philosopher, Professor Agassiz, in explanation of the above facts, says, "that an *immaterial principle*, which no external influence can prevent or modify, is present, and determines its future form, so that the egg of a hen can produce only a chicken, and the egg of a codfish only a cod."

man and the higher animals, especially the anthropomorphous apes. Having cited various authorities to prove the truth of the above statements, he finishes his introductory chapter by saying, that time will before long come, when it will be thought wonderful that naturalists, who were well acquainted with the comparative structure of man and other mammals, should have believed that each was the work of a separate act of creation.

Having shown that there is no essential difference between man and the higher mammals in their corporeal organisation, he then passes on to the consideration of the mental qualities, where, of course, a much wider gulf would be expected to exist; and even here, he points out that the germs of all our intellectual characteristics, and some of our moral, are to be found among the lower animals. He argues that man and the higher animals, especially the primates, have the same senses, intuitions, and sensations; similar passions, affections, and emotions; that they feel wonder and curiosity; that they possess the same faculties of imitation, attention, memory, love, imagination, and even reason, though in different degrees. Having admitted that this difference is enormous—even if we compare the mind of one of the lowest savages, who has no words to express any number higher than four, and who uses no abstract terms for the commonest objects or affections, with that of the most highly organised ape—he insists, nevertheless, that the difference in mind between man and the higher animals, great as it is, is certainly one of *degree and not of kind*.

The main conclusion arrived at by Mr. Darwin is, that man is descended from some lowly-organised form, and that "with all his noble qualities, with sympathy which feels for the most debased, with benevolence which extends not only to other men but to the humblest living creature, with his god-like intellect which has penetrated into the movements and constitution of the solar system—with all these exalted powers—Man still bears in his bodily frame the indelible stamp of his lowly origin."*

I wish here to make a brief comment upon a most able notice of the "Descent of Man," which appeared in the *British Quarterly Review* for October, 1871. Agreeing as I do with the general tenor of the writer's remarks, I most entirely differ

* " Descent of Man," Vol. II., P. 405.

from him in one essential point. After disputing the truth of Mr. Darwin's assumed similarity between the minute structure of man and animals, he goes on to say, " If it could be shown that in their minute anatomy the tissues of an ape so closely resembled those of a dog on the one hand, and of a man on the other, as that they could not be distinguished by the microscope, the fact would be of the highest importance, and would add enormously to the evidence already adduced by Mr. Darwin." I cannot agree with the inference here drawn by the able reviewer, who seems to imply that Mr. Darwin's theory is unassailable if he can prove his assertion as to the close similarity in the minute structure of man and animals. I am ready to admit this similarity; I will even strengthen Mr. Darwin's position by admitting that there is a remarkable correspondence in the vital properties of the blood of man and animals, as shown by the fact that in the case of apparent death in man from loss of blood, resuscitation has taken place in consequence of the transfusion into the system of the blood of an animal, as the sheep or the calf.* It is

* This analogy, however, in the vital properties of the blood must not be supposed to imply identity in the chemical composition. On the contrary, the microscope and chemical analysis have shown not only that the blood of man differs from that of the lower animals, but that the blood of each species of animal differs from that of every other species. It is stated even in our modern treatises on Medical Jurisprudence, that the microscope can merely determine whether blood is derived from the class Mammalia, or from a bird, fish, or reptile; but an American writer, Dr. J. C. Richardson, in an able and elaborate forensic essay on the diagnosis of blood-stains, has recently shown that the red blood-discs of animals with rounded corpuscles, are just as distinct in different animals as are different kinds of shot, and that we are now able, by the aid of high

idle to attempt to shirk the import of these physiological results. I admit the force of them. I do not deny that man is an animal, and that he has the essential properties of a highly organised one; he is constructed on the same general type or model as other mammals. All vertebrate animals have many characteristics in common, chemical composition, cellular structure, laws of reproduction, growth, decay, and death; and the resemblance may even be extended to the Brain, where

powers of the microscope, and under favourable circumstances, to positively distinguish stains produced by human blood from those caused by the blood of various other animals, and this even after the lapse of five years from the date of their primary production! The facts upon which these statements are founded are fully discussed in the *British Quarterly* for October, 1871, and in the *American Journal of Medical Sciences* for July, 1874, to which periodicals I would refer the reader for much valuable information upon this important subject.

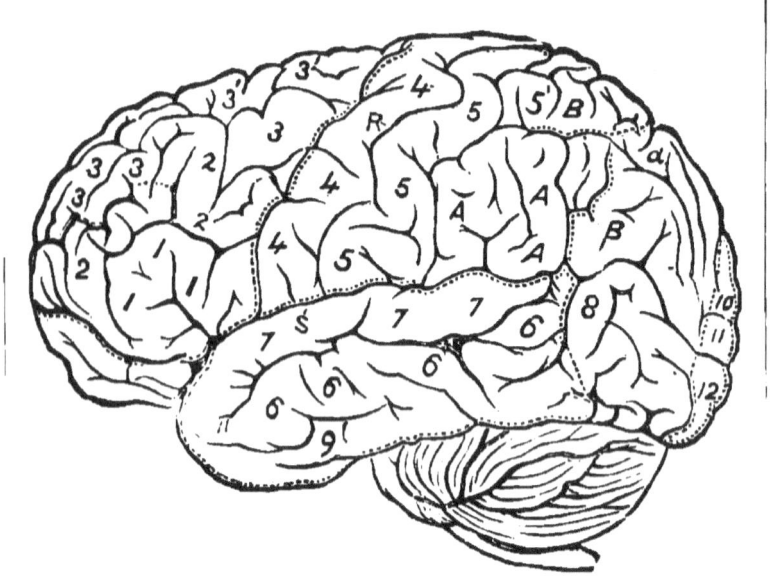

Fig. VII.—Brain of Man *(Homo)*.

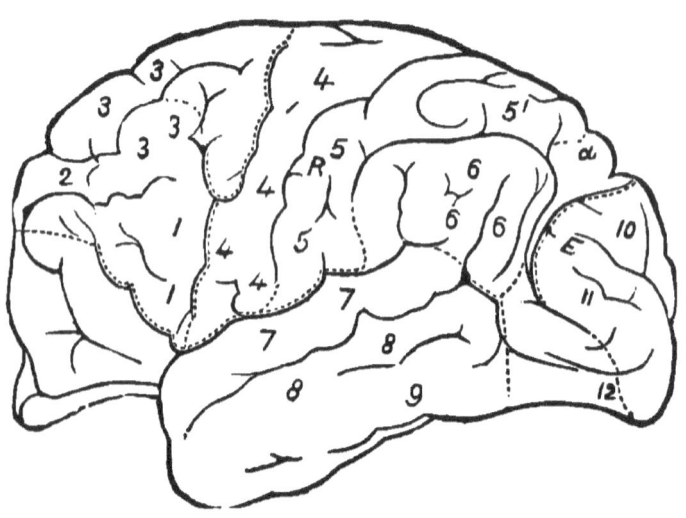

Fig. VIII.—Brain of Orang *(Simia)*.

every chief fissure and fold in Man has its analogy in the Orang, the Chimpanzee, and the Gorilla, as may be seen by comparing the Brain of Man with that of the Orang, as represented in Figures 7 and 8. I admit all this, and I agree with Hallam that "the framework of the body of him who has weighed the stars, and made the lightning his slave, approaches to that of the speechless brute that wanders in the forests of Sumatra."

Whilst, however, conceding that Man, in his purely physical nature, is closely allied to certain members of the brute creation, I entirely repudiate the inference drawn from this analogy by Mr. Darwin and other writers of the modern school of thought; for supposing it to be proved to a mathematical demonstration, that Man is like an Ape, bone for bone, muscle for muscle, nerve for nerve, what then? What does this prove, if it can be shown that Man possesses a *distinctive attribute*, of which not a trace can be found in the Ape,—an attribute of such a nature as to create an immeasurable gulf between the two? This attribute I assert to be the faculty of Articulate Language, which I maintain to be a difference, *not only of degree, but of kind.*

# CHAPTER IV.

Τὰ δὲ ζωοτόκα καὶ τετράποδα ζῶα ἄλλο ἄλλην
ἀφίησι φωνήν, διάλεκτον δ' οὐδὲν ἔχει, ἀλλ'
ἴδιον τοῦτ' ἀνθρώπου ἐστίν.

ARISTOT. *Hist. Anim.* iv. 9. 16.

*Definition of Language—Stoddart, Trench, Whitney, and Farrar—Language, a Distinctive Attribute of Man—The so-called language of the Parrot considered—Connection between Ideas and Words—Cases of Heterophasia.*

Having been engaged for some years past in studying the question of the localisation of the Faculty of Speech, and believing that my published researches furnish a powerful and original argument against the doctrine of evolution, I trust I may, without presumption, be allowed to indulge the hope that I can furnish an additional and original argument against this dangerous heresy, by showing that the possession of Articulate Language establishes a difference between man and animals—a difference not of degree only, but of kind.

In order to establish my position, I shall have to consider three propositions:—

I.—That Articulate Speech is a *Distinctive* Attribute of Man, and that the Ape and lower animals do not possess a trace of it.

II.—That Articulate Speech is a *Universal* Attribute of Man—that all races have a Language, or the capacity of acquiring it.

III.—The Immateriality of the Faculty of Speech.

## DEFINITION OF LANGUAGE.

I now propose very briefly to explain what I understand by the term Faculty of Language, in reference to which so much ambiguity and confusion exist, as to render a clear definition absolutely necessary. I shall then inquire how far this faculty is shared by animals, and having shown that they do not possess it even in an elementary form, I shall then glance at the much-disputed question of the Seat of Language—the Localisation of the Faculty of Speech,—a question which seems to me to have an important bearing on the point at issue.

Of all the branches of knowledge, there are none more interesting than the study of language, that marvellous faculty of expressing human thought, and which is included in the gift of reason to man. Speech, the expression of the conceptions of the mind by articulate sounds, is one of the most valuable possessions which adorn and elevate our being, it is the instrument of our thoughts, the organ of our social nature, and the most important means of our communication with our fellow-men. 'It enables us,' says Stoddart, 'as it were, to express things beyond the reach of expression, the infinite range of existence, the exquisite fineness of emotions, the intricate subtleties of thought. Of such effect are those shadows of the soul, those living sounds, which we call *words!* Compared with them, how poor are all other monuments of human power, or perseverance, or skill, or genius! They render the mere clown an artist; nations immortal; orators, poets, philosophers, divine!"*

\* The Philosophy of Language," by Sir John Stoddart. P. 1.

# LANGUAGE DEFINED. 91

'Language is the embodiment, the incarnation of the feelings and thoughts and experience of a nation; it is the amber in which a thousand precious and subtle thoughts have been safely embedded and preserved. It has arrested ten thousand lightning flashes of genius, which unless thus fixed and arrested, might have been as bright, but would have also been as quickly passing and perishing as the lightning.'*

'It is necessary to bear in mind,' says Dr. Carpenter, 'that Vocal sounds and Speech or articulate language are two things entirely different; and that the former may be produced in great perfection, where there is no capability for the latter. Hence we should infer that the instrument for the production of vocal

* "Trench on the Study of Words," P. 23.

sounds was distinct from that by which these sounds are modified into articulate speech; and this we easily discover to be the case, the voice being unquestionably produced in the larynx, whilst the modifications of it by which language is formed, are effected for the most part in the oral cavity.'*

Man shares with animals the power of emitting sounds by means of an apparatus especially adapted for that purpose; sound being described as a particular movement of ponderable matter capable of affecting the organ of hearing. 'Each one of the sounds composing our spoken alphabet,' says Professor Whitney, 'is produced by an effort in which the lungs, the throat, and the organs of the mouth bear a part. The lungs furnish the rough

* "Principles of Human Physiology," P. 958.

material, an expulsion of air, in greater or less force; the vocal cords in the larynx by their approximation and vibration give to this material resonance and tone; while it receives its final form, its articulate character, by the modifying action of the tongue, palate, and lips. Each articulation thus represents a certain position of the shaping organs of the mouth, through which a certain kind and amount of material is emitted. A word is composed of a series of such articulations, and implies a succession of changes of position in the mouth-organs, often accompanied by changes in the action of the larynx upon the passing column of air. A spoken alphabet is no chaos, but an orderly system of articulations, with ties of relationship running through it in every direction.'* Man alone possesses the power of regulating and systematising these sounds, so as to transmit to others the impressions of his mind in the form of a language, which has been described as a sensible phenomenon by which thought becomes materialised. 'The essence of speech,' says Dean Goulburn, 'is not in the sound, otherwise a machine might be made to speak.' In fact, speech or language consists of a series of conventional sounds, which represent a meaning which the mind has previously attached to their expression; it is, in fact, the power of connecting definite sounds with definite ideas, thus constituting a medium by which ideas are conveyed from mind to mind in logical method. 'The essence of language,' says

* "Language and the Study of Language," by W. D. Whitney, Professor in Yale College. Pp. 87, 91.

an able writer in the *Quarterly Review*, 'is mental—an intellectual activity called the *verbum mentale*; but actual 'speech' itself is the outward expression of thoughts (rational conceptions) by articulate sounds —the *verbum oris*. We may have (1) animal sounds that are neither rational nor articulate; (2) sounds that are articulate but not rational; (3) sounds that are rational but not articulate; (4) sounds that are both rational and articulate; (5) gestures which do not answer to rational conceptions; and (6) gestures which do answer to such conceptions, and are, therefore, external but non-oral manifestations of the verbum mentale. The fourth category is that of true speech.'\*
According to a French psychologist, M. Parchappe, the exercise of the function

\* "Quarterly Review," July, 1874.

of speech is accomplished by the manifestation of three distinct modes of physical force; development of *intellectual force* in the formation of an idea; of *voluntary force* in the determination of acts necessary to translate this idea into words; and, lastly, of *motor force* in the voluntary manifestation of the movements resulting in articulate voice. In short, according to M. Parchappe, the function of speech comprises three essential physical elements: Intelligence, Volition, and Movement. 'The voice,' says Dr. Farrar, 'is the organ of the understanding; and speech is the expression of the thinking spirit in articulate sounds—the union of sound and sense, the combination of the phonetic and the intellectual elements into one organic unity.' *

* " Chapters on Language," P. 84.

There are two distinct features in speech,—an act of the intelligence, and a sonorous mechanism. These have been termed *cognitive* and *executive*,—thought-speech and spoken-speech; the internal and external speech of M. Bouillaud. The latter is what Lord Monboddo calls 'the *material part* of language; for of the breath modified by the organs of the mouth, is produced articulation; and the mind furnishes the ideas, which make the form of language.' \*

Here I would remark that it is important not to confound the faculty of *articulate* language with the *general* faculty of language, and Professor Broca's remarks on this subject are so lucid and terse that I cannot do better than transcribe them :—

\* "On the Origin and Progress of Language," vol. ii., P. 3.

'There are several kinds of language; every system of signs which permits the expression of ideas in a manner more or less intelligible, more or less complete, or more or less rapid, is a language in the general sense of the word: thus speech, mimicry, dactylology, writing both hieroglyphic and phonetic, are so many kinds of language. There is a general faculty of language which presides over all these modes of expression of thought, and which may be defined, the faculty of establishing a constant relation between an idea and a sign, be this sign a sound, a gesture, a figure, or a drawing of any kind.'[*]

In order to establish my first proposition, we must now inquire whether

---

[*] "Sur le Siége de la Faculté du Langage Articulé," P. 4.

language is the exclusive prerogative of man? Some would answer this question in the negative, and a celebrated French anthropologist, M. Coudereau, maintains that man is not alone in possessing a language; that all species of animals possess one, varied, but sufficient to express their ideas. He further says that 'man acquires the faculty of speech by his memory, labour, and imitation,— the parrot does no more. From a linguistic stand-point, this faculty is in its nature identical in man and animals; man can articulate sounds, other animals can imitate sounds as well as he can. He presents simply, in this respect, a greater development of a faculty common to all social animals.'

M. Lemoine, in a highly philosophical treatise, devotes a chapter to the Language of Animals, in which he says 'no animal speaks, but all, or nearly all, have special signs of their own; they cry or they sing, these cries or songs varying according to the passions by which the animals are influenced, and they are understood naturally by all individuals of the same species. These modes of expression have neither been learnt from their parents, whom they have often not known, nor have they been acquired by experience, since they are not more developed in advanced age than in youth, and, moreover, this language of animals is the same now as it was in the time of Pliny or of Aristotle. Animals have the peculiarity of the cry, man has the peculiarity of speech. Speech is as natural to man as the cry or the song is to animals.' *

\* " La Physionomie et La Parole." Paris, 1865.

## LINGUISTIC MYTHOLOGY. 101

Of course, in a matter of this kind, no importance whatever can be attached to evidence obtained from the Greek and Latin classics, but, as a matter of curiosity, I may mention that Homer represents Xanthus, the horse of Achilles, as having been rewarded by Juno with the gift of speech.* Livy also informs us that an ox once uttered these words, *Roma cave tibi*.† Such stories as these, of course, need no serious refutation.

Professor Max Müller's evidence upon this point is given with no uncertain sound. 'However much,' says he, 'the frontiers of the animal kingdom have been pushed forward, so that at one time the line of demarcation between animal and man seemed to depend on a mere

* Iliad. xix, 405.
† Liv: Lib. xxxv. Cap. 21.

fold in the brain, there is *one* barrier which no one has yet ventured to touch —the barrier of language. The faculty of speech is the distinctive character of mankind, unattained and unattainable by the mute creation. It distinguishes man from all other creatures; and if we wish to acquire more definite ideas as to the real nature of human speech, all we can do is to compare man with those animals that seem to come nearest to him, and thus try to discover what he shares in common with these animals and what is peculiar to him, and to him alone.' * In a later publication, the same writer observes, 'there is between the whole animal kingdom on one side, and man, even in his lowest state, on the other, a

* "Lectures on the Science of Language," Pp. 14, 383, 385.

barrier which no animal has ever crossed and that barrier is—*Language*. By no effort of the understanding, by no stretch of the imagination, can I explain to myself how language could have grown out of anything which animals possess, even if we granted them millions of years for that purpose. If anything has a right to the name of *specific difference*, it is language, as we find it in man, and in man only. Even if we removed the name of specific difference from our philosophic dictionaries, I should still hold that nothing deserves the name of man except what is able to speak.'*

The next authority I wish to quote is Stuart Mill, who, writing on the same subject, says 'the attribute of being

* " Lectures on Mr. Darwin's Philosophy of Language, Fraser's Magazine," June, 1873.

capable of understanding Language is a Proprium of the species man, since, without being connoted by the word, it follows from an attribute which the word does connote, viz., from the attribute of rationality.'*

Mr. Darwin, whilst admitting that language has justly been considered as one of the chief distinctions between man and the lower animals, adds, however, that man uses, in common with the lower animals, inarticulate cries to express his meaning, aided by gestures and the movement of the muscles of the face, and he doubts not 'that language owes its origin to the imitation and modification, aided by signs and gestures, of various natural sounds, the voices of other animals, and man's instinctive cries.' He suggests the

* "A System of Logic," Vol. i. P. 180.

probability that 'primeval man, or rather some early progenitor of man, used his voice largely, as does one of the gibbon apes at the present day, in producing true musical cadences—that is, singing;' and it does not appear to him altogether incredible, that 'some unusually wise ape-like animal should have thought of imitating the growl of a beast of prey, so as to indicate to his fellow-monkeys the nature of the expected danger; and this would have been a first step in the formation of a language.' * A writer in the *Edinburgh Review*, commenting upon the above passage, asks for the evidence

* Man's power to construct a language for himself has been called the *Bow-wow* theory and the *Pooh-pooh* theory, or the Onomatopoetic and Interjectional theories. Professor Max Müller, in his Lectures on the Science of Language, shows the untenability of this doctrine, and, speaking of the Bow-wow theory, says,

H

that at the present day some unusually wise ape has ever been known to imitate the cry of a wild beast, so as to indicate its presence to its fellows. 'Why, also, if the first stage of articulate development began in musical cadences, by which the chords of the voice were strengthened and gradually perfected, and if the second consisted in the imitation of other sounds, have not the birds evolved for themselves an articulate language, seeing that they exercise their voices at least as much as any of the higher animals.'\* Mr. Darwin

'it goes very smoothly as long as it deals with cackling hens and quacking ducks; but round that poultry-yard there is a dead wall, and we soon find that it is behind that wall that language really begins.' "Lectures on the Science of Language," Second Series, P. 91.

\* Mr. Lyon in combating Mr. Darwin's linguistic theory, observes, that 'Mr. Darwin takes for granted what he cannot prove—viz., that man had ape-like progenitors, and that some one of them possessed mental says that the sounds uttered by birds offer in several respects the nearest analogy to language, and he lays great stress upon

powers more highly developed than those of any existing ape. Reasoning from this highly-developed, hypothetical ape, he tells us that, by exercising what power of utterance it had, the brain enlarged and the mind improved, and the vocal organs strengthened, generation after generation, till this series of changes in a race of apes culminated in man! But all this is purely imaginary. Mr. Darwin cannot produce even the shadow of a proof that this 'unusually wise ape-like animal' ever existed to transmit his wisdom to his descendants, or that he had descendants to inherit it, yet he tells us we may 'confidently believe' it! Instead of trying to prove to us that such development has occurred, he asks us 'confidently to believe' that it has occurred! It is a singular circumstance, moreover, that, if the 'unusually wise ape-like animal' which he supposes took the first step in the formation of a language, ever really existed, there should not have arisen other 'unusually wise' apes to take further steps in the same direction, so that there should have been speaking apes in the present day. But no existing race of apes seems to have got beyond the *growl* of which Mr. Darwin has spoken.' "Homo versus Darwin," P. 106.

the fact that parrots can talk. Now, I maintain that the so-called talking of the parrot is not articulate language, it is merely the result of a remarkable power of imitation possessed by that bird, which faculty of imitation can exist in the human subject after the power of language has ceased. The following case observed by myself will illustrate my meaning :— During a recent visit to La Salpêtrière, an institution in Paris for the reception of female patients, for the most part afflicted with some mental disorder, the physician, Dr. Auguste Voisin, knowing I was interested in the question of language, called my attention to the case of an old woman in whom the faculty of speech was completely suspended, but, who, although she never spoke, repeated like a parrot all that was said before her. For instance, Dr.

Voisin addressed her thus:—'Voulez-vous manger aujourd'hui?' She said instantly, 'Voulez-vous manger aujourd'hui?' I then said to her, 'Quel âge avez-vous?' She replied, 'Quel âge avez-vous?' I then said to her in English, 'You are a bad woman.' She instantly replied, 'You are a bad woman.' I said, 'Sprechen sie Deutsch?' She retorted, 'Sprechen sie Deutsch?' In the words that she thus echoed, her articulation was distinct, although the foreign phrases were not repeated by her in quite so intelligible a manner as the French. Not only did this woman echo all that was said, but she imitated every gesture of those around her. One of the pupils made a grimace; she instantly distorted her facial lineaments in precisely the same manner. Another pupil made the peculiar defiant action, common in schoolboys, of putting the thumb to the nose and extending all the fingers, called in French, *pied de nez*. The patient instantly imitated this elegant performance. Just as we were leaving her bedside, a patient in an adjoining bed coughed; the cough was instantly imitated by this human parrot! In fact, this singular old woman repeated everything that was said to her, whether in an interrogative form or not; and she imitated every act that was done before her, and that with the most extraordinary exactitude and precision.

I have mentioned this case to show that the faculty of imitation seems to be independent of that of speech. The parrot may be taught automatically to do, in an imperfect degree, what this old woman did but that does not imply the possession of language.

I would ask of those gentlemen who attach so much importance to pantomimic expression, and to the power of imitation possessed by certain animals, why it is that, under the influence of domestication, no monkey or parrot has ever evolved for itself an articulate language? The parrot probably possessed the same power of imitation 3,000 years ago, and yet we see no probability of its gradual development into a more decided form of expression; the monkey, too, whose structural organism so closely resembles that of man, has never evinced the slightest aptitude for the acquisition of Articulate Language. I believe with Max Müller, that " speech is the one great barrier between the brute and man, and that no process of natural selection will ever distil significant words out of the notes of birds or the cries of beasts. Language is our Rubicon, and no brute will dare to pass it."

I must now proceed to consider a point which has a collateral connection with my subject, and to answer a question raised by the Dean of Norwich in his work entitled 'The Idle Word,' a book in which I have met with much to corroborate my views as to the *immateriality* of the Faculty of Speech; in fact, it seems that the Dean's thoughts have sometimes run in the same mental grove as mine. At page 17, he says, 'It is a very old debate whether or not it is possible to reason mentally, without having the words in the mind, which represent the subjects of our reasoning.' Now, I can answer this question affirmatively, as will be seen by the following cases of perversion of speech which have been recorded by two of the most distinguished physicians of modern times :—Dr. W. D. Moore, of Dublin, had under his care a gentleman, who, although his intelligence was unimpaired, had completely lost the connection between ideas and words. On one occasion, Dr. Moore was much puzzled by his patient, who was in bed, saying to him, *" Clean my boots."* Finding that he was not understood, he became much excited, and cried out vehemently *" Clean my boots by walking on them !"* At length it was ascertained that the cause of his disquietude was the shining of the candle on his face, and that the object of his unintelligible sentences was to have the curtain drawn; when this was done he appeared quite gratified. The subject of his reasoning was the *drawing of the curtain*, but the words used were, *" Clean my boots."*

Another still more remarkable instance of the want of connection between words employed and the ideas intended to be conveyed, is recorded by the late Professor Trousseau, of Paris, the subject of it being a lady, Madame B——, the mother-in-law of a physician, who was affected with the following strange misapplication of language :—Whenever she received a call from a visitor, she rose to receive him with a benevolent smile on her countenance, and pointing to a chair, said—*" Pig, Brute, Stupid Fool."* Madame B—— begs you to be seated, her son-in-law would then say, giving this interpretation to her wishes thus strangely expressed. Here, again, the idea in this lady's mind was courteously to ask her visitor to be seated, whilst the words actually used were those of coarse and vulgar abuse.

These instances of perversion of language, to which the name of Heterophasia has been appropriately given, conclusively show that it is possible to reason mentally, without having the words in the mind, which represent the subjects of our reasoning.

# CHAPTER V.

'Ο δὲ νοῦς ἔοικεν ἐγγίνεσθαι οὐσία
τις οὖσα, καὶ οὐ φθείρεσθαι.
ARISTOT. *De Anima*. I. 4. 12.

*The Anatomical Seat of Speech—Rôle of the Cerebral Convolutions; Flourens, Maudsley—Gall's Phrenological System—Destruction of the anterior lobes of the brain without impairment of the power of speech—Comparative development of the third frontal convolution in Man and in the Ape—Speech is a barrier the brute is not destined to pass.*

Having defined what is meant by the faculty of Language, I now proceed to review very briefly the various theories which have been from time to time promulgated as to the Seat of Articulate Language, as the question of the localisation of this faculty seems to me to have an important bearing upon the point at issue ;* but, before doing this, it is imperative that I should enter into a few anatomical details for the better understanding of my subject, as I am justified in assuming that a portion of my readers may be but imperfectly acquainted with the main divisions of the brain.

The encephalon is a collective term,

---

* The cerebral localisation of language has of late years engrossed the attention of physiologists in all parts of the world; and, in this country, an additional stimulus has recently been given to this inquiry by the interesting experiments of Professor Ferrier on the localised application of electricity to the surface of the brain. The subject is so vast, that anything beyond a mere allusion to it would be beyond the scope of this essay. The comparative value of the various theories as to the Seat of Speech are fully discussed in the author's treatise "On Aphasia, or Loss of Speech, and the Localisation of the Faculty of Articulate Language," to which work he would refer those of his readers who may desire more detailed information upon this obscure and much controverted subject.

which signifies those parts of the nervous system which are contained within the cranium, viz., the cerebrum, or brain proper, the cerebellum, and the medulla oblongata. The cerebrum is by far the largest portion of the encephalon, and consists of two lateral halves called *hemispheres*, each hemisphere being subdivided into three *lobes*,—anterior, middle, and posterior. The hemispheres present upon their surface numerous smooth and tortuous eminences called *convolutions*, which have received special names, those only which concern my subject being the frontal convolutions, which are known as first, second, and third frontal. It has been maintained that man's intellectual superiority is principally due to the depth and extent of the cerebral convolutions, which are wanting in all classes below the Mammalia, and they are absent even in the lower members of this class. According to Flourens, the Rodentia, the least intelligent of the Mammalia, have no convolutions; the Ruminantia, more intelligent than the Rodentia, possess them; the Pachydermata, who are still more intelligent than the Ruminantia, have still more convolutions; and so on the number continues to increase as we ascend to the Carnivora, then to the Apes, the Orangs, and lastly to Man, who is the richest of all animals in cerebral convolutions. If this gradation in the number of the convolutions have a relation to the intelligence of the animals, it would seem to give an *a priori* reason for concluding that the manifestation of the highest product of intelligence—speech—may well have some connection with the development of the convolutional grey matter.*

Of the cerebellum I need say nothing,— it has no reference to the subject of my remarks.

The medulla oblongata is that part of the encephalon which is placed immediately above the spinal cord, forming the bond of union between it and the brain. It is divided into two lateral columns, which are themselves subdivided into three smaller cords, called the pyramidal, olivary, and restiform bodies.

---

* One of our leading psychologists, Dr. Maudsley, says that 'we cannot at present exhibit an exact relation between the development of the convolutions and the degree of intelligence in different animals; for the brains of the ass, the sheep, and the ox are more convoluted than those of the beaver, the cat, and the dog; but the relative size of the animal must be taken into consideration in such comparison.'

The ancients seem to have possessed the most crude notions of the functions of the brain, as evidenced by Hippocrates assigning the seat of the mind to the left ventricle, and also by Aristotle placing the sensorium commune in the heart, the brain, according to him, being an inert viscus bloodless and cold, serving only as a refrigerator to the heart. Michael Servetus, who flourished in the sixteenth century, believed the choroid plexus was the organ destined to secrete the animal spirits, that the fourth ventricle was the seat of memory, and that the habitation of the soul was in the aqueduct of Sylvius; a century later, René Descartes assigned to the soul a more secure position in the pineal gland, from which, however, it was soon dislodged by our fellow-countryman, Thomas Willis, who disputed its right to this central spot, on the ground that 'animals which seem to be almost destitute of imagination, memory, and other powers of the soul, have this gland large and fair enough.'*

In later times the brain has been universally considered to be the organ of thought and intelligence; but opinions have been, and are still, divided as to whether it is to be regarded as a single organ, or as consisting of a series of distinct organs, each endowed with a special and independent function; whether, in fact, the phenomena of intelligence are due to an action of the brain as a whole, or whether the different psychological elements which constitute them are connected with isolated and circumscribed

---

* Cerebri Anatome cui accessit nervorum descriptio, Cap. xiv., P. 102, (1667.)

parts of the encephalon.* Out of this last theory has arisen the principle of the localisation of the cerebral faculties, which was, in the early part of the 19th century, announced in a definite form by Gall, who divided the brain into organs endowed with primordial faculties, distinct the one from the other. The germ of this idea of the polysection of the encephalon is to be found in the writings of physiologists long before the time of Gall; indeed, one author, Charles Bonnet, assigned a special function to each fibre, stating that every faculty, sensitive, moral, or intellectual, was in the brain connected to a bundle of

* All are agreed, says Dr. Ferrier, 'that it is with the brain that we feel, and think, and will; but, whether there are certain parts devoted to particular manifestations, is a subject on which we have only imperfect speculations, or data too insufficient for the formation of a scientific opinion.'

fibres; that every faculty had its own laws which subordinated it to other faculties, and determined its mode of action; and that not only had every faculty its fasciculus of fibres, but that every word had its own fibre!* This writer is not very logical in his conclusions, for he maintains that each brain has, from the birth of the individual, characters which distinguish it from every other brain; and after stating that 'it is as impossible for a passionate man to be otherwise than passionate, as it is for the three angles of a triangle to be otherwise than equal to two right angles,' he utterly destroys the force of his reasoning by the following passage:—'Whence comes

* A Spanish physiologist, Juan Huarte, writing in the sixteenth century, proposed that a jury of scientific men should determine what course of study, and what career should be assigned to each child.

the enormous distance which separates the immortal Newton from the rustic peasant? Has nature not moulded their brains out of the same material? Has she, perchance, placed in one of these brains certain parts which are not to be found in the other? Or, has she arranged these parts in a different manner in each brain? No, the brain of the peasant has essentially the same organs, the same structure, and the same texture as the brain of the philosopher. Education alone has effected this prodigy.'*

Gall, however, was the first to attempt to connect the seat of language with any definite portion of the cerebro-spinal centre, by asserting that there was a special organ for language, which, according to him, was placed in those convolutions of the

* Essai de l'sychologie, P. 159. (1754.)

anterior lobes of the brain, which rest upon the posterior part of the supra-orbital plates, or, in other words, upon the roof of the orbit. These convolutions are marked O, O, in Figure IX, which is a representation of the convex surface of the left hemisphere, the engraving being taken from a cast kindly sent to me by my friend Professor Broca, of Paris.

The circumstance which directed Gall's attention to the possibility of connecting the brain with certain faculties of our mental nature is so well known that I scarcely need to allude to it. In his early days, he often found himself surpassed by certain of his fellow-students who he felt were intellectually inferior to himself, but in whom a remarkable memory coincided with a striking prominence of the ocular globes. This external prominence

# FIG. IX.—CONVEX SURFACE OF THE LEFT HEMISPHERE,

### Showing the Disposition and Arrangement of the Cerebral Convolutions.

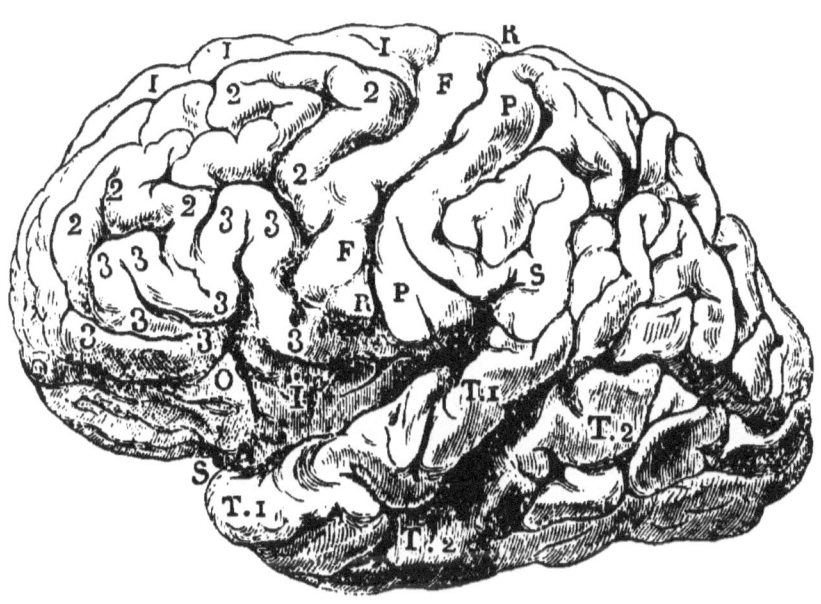

RR, Fissure of Rolando.   SS, Fissure of Sylvius.
1, 2, First and second frontal convolutions.
3, Third frontal convolution, in the posterior part of which M. Broca places the seat of Speech.
FF, Transverse frontal convolution.
PP, Transverse parietal convolution.
OO, Orbital convolutions, the seat of language according to Gall.
T1, T2, First and second temporo-sphenoidal convolutions.
I, Island of Reil (the superior and inferior marginal convolutions are represented as being drawn asunder so as to expose it).

led him to the inference that there was an internal cerebral prominence which produced it, and it was the application of this reasoning to other cranial protuberances that gave rise to his craniological doctrine.

This is not the place to make more than a passing allusion to Gall's views, as they have not met with anything like general acceptance. He was not, however, altogether without admirers amongst the scientific celebrities of his day, one of the most enthusiastic of whom was the French physiologist, Broussais, who, on the organ of murder being found in the sheep, attempted to reconcile this fact with Gall's doctrine, by asserting that the destruction of vegetables might be compared with the destruction of animals! At Rome, the Pope paid the same compliment to Gall as his predecessor had done to Copernicus; and a sort of panic was occasioned at Vienna, by Gall's desire to become possessed of the skulls of deceased Austrian celebrities, and it is said that the Emperor's librarian added a codicil to his will, enjoining that his skull should not be delivered up to the profanation of this modern Democritus.

Gall's labours would undoubtedly have met with a more hearty recognition from his contemporaries, had not the Austrian priesthood raised the cry of *materialism* as applied to his doctrines. The great German psychologist had no such heterodox notions as his adversaries maliciously attributed to him,* for, as

---

\* Let us hear Gall himself upon this point, 'When I say that the exercise of our moral and intellectual faculties depends on material conditions, I do not mean

Hufeland philosophically observes, 'he was employed in analysing the dust of the earth of which man is formed, not the breath of life which was breathed into his nostrils.'

As in Gall's days so in ours, this very indefinite and unmeaning word '*materialism*,' is used as a kind of psychological scarecrow, to frighten all those who are endeavouring to trace the connection between matter and mind. Surely there is nothing contrary to sound theology, in assigning certain attributes or functions of an intellectual order to certain parts of our nervous centre ; the same power that

that our faculties are the *product* of the organisation; this would be confounding *conditions* with *efficient causes*. I limit myself to what can be submitted to our observation'—" *Sur l'origine des qualités morales, et des facultés intellectuelles de l'homme,*" par F. J. Gall. Tom. I., P. 189.

made the sun, the 'powerful king of day' that early morn sees rising from ocean's billowy bed, that made the stars those countless orbs of light that gem the vaulted sky—this same power, surely, could just as well ordain that a multiplicity of organs should be necessary to the full development of man's mental faculties, as that the manifestation of them should depend upon the integrity of one single organ; the cerebral localisation of our divers faculties, and the plurality of our cerebral organs, strike no blow at the great principle of the moral unity of man.

Gall's conclusions must be considered in many instances arbitrary and hypothetical; still, I would say, let not the spark be lost in the flame it has served to kindle, for, in spite of all that has been said against Gall, and all that has been written in depreciation of his labours, beyond all doubt his researches gave an impulse to the cerebral localisation of our faculties, the effect of which is especially visible in our own days; and although his great work on the "Functions of the Brain" was received with sneers, scoffs, and ridicule by his contemporaries, I look upon it as an imperishable monument to the genius and industry of one of the greatest philosophers of the present age.

Although not the next theory in chronological order, it is convenient here to make a passing allusion to the views of a Dutch physiologist, Professor Schrœder Van Der Kolk, who placed the seat of speech in the olivary bodies. Besides citing numerous cases in illustration of his hypothesis, he gave an *a priori* reason for his theory in the fact, that the olivary bodies occur only in mammalia; that, on comparing these organs as occurring in mammals themselves, they are most developed in man; and that in the higher mammalia, as the ape, they are most like those in man. This hypothesis, which has never met with much support, has been rejected by most physiologists of the present day.

I now arrive at the consideration of certain theories which have a more direct reference to my subject,—I mean those which locate speech in the anterior lobes of the brain, or in some particular fold of these lobes.

As far back as 1825, Professor Bouillaud, of Paris, placed the faculty of articulation in the anterior lobes of the brain, which

he considered to be the organs of the
formation of words and of memory; and he
stated that the exercise of thought de-
manded the integrity of these lobes. He
supported his position by reference to 114
cases in which loss or impairment of speech
coincided with disease of the anterior
lobes.* Such was M. Bouillaud's confidence in his theory, that he offered a prize

* M. Bouillaud did not confine his pathological
investigations to the human subject, but instituted a
series of experiments upon animals, with the view of
determining the functions of the brain, and on several
occasions he removed different portions of the cerebral
lobes, without impairing sight or hearing ; he also
removed the entire hemispheres from a chicken, in
whom the power of expressing pain by its peculiar cry
was retained. On one occasion, he pierced with a
gimlet the anterior part of the brain of a dog, from
side to side, at a spot corresponding to the union of
the anterior with the middle lobes—that is in the
immediate neighbourhood of Broca's region. The dog
survived the mutilation, but was much less intelligent
than before the operation, and although he could utter

of 500 francs for any well-authenticated case in which the two anterior lobes were destroyed, or more or less seriously injured, without speech being affected. This challenge remained unaccepted for many years, till the occurrence of a celebrated discussion on the seat of language, at the Academy of Medicine of Paris,* when M. Velpeau said he should

cries of pain, he had entirely lost the power of barking. As far as the present inquiry is concerned, I am aware that but little importance can be attached to these experiments, for there is little or no analogy between the cry of a chicken or the bark of a dog, and the articulate speech of man; still, experiments of this kind may have an indirect bearing upon our subject, and it would be extremely interesting to know what would be the effect of traumatic injury to certain regions of the anterior lobes of the brain, upon the quasi-articulatory powers of the parrot.

* For a detailed account of this memorable debate, which extended over several meetings, vide "Bulletins de l'Académie de Médecine." Tom. xxx.

claim the prize on the faith of the following case observed by himself:—

'In the month of March, 1843, a barber, sixty years of age, came under M. Velpeau's care for disease of the prostate gland. With the exception of his prostatic disorder, he seemed to be in excellent health, was very lively, cheerful, full of repartee, and evidently in possession of all his faculties; one remarkable symptom in his case being his *intolerable loquacity*. *A greater chatterer never existed*, and on more than one occasion complaints were made by the other patients of this talkative neighbour, who allowed them rest neither night nor day. A few days after admission this man died suddenly, and a careful autopsy was made. On opening the cranium, a cancerous tumour was found, which had taken the place of the two anterior lobes!'

Here then was a man, who, up to the time of his death, presented no symptom whatever of cerebal disease; who, far from having any lesion of the faculty of speech, was *unusually loquacious;* and who, for a long period prior to his decease, must have had a most grave disease of the brain, which had destroyed a great part of the anterior lobes. Surely this case alone, recorded by such a high authority as M. Velpeau, ought to be sufficient utterly to subvert the theory of the localisation of speech in the anterior lobes; but I have still further evidence to adduce. M. Peter has recorded the case of a man who fractured his skull by a fall from a horse; after recovery from the initial stupor, there succeeded a *remarkable loquacity*, although after death it was found that the two frontal lobes of the brain were reduced to a pulp (*réduits en bouillie*). Again, Professor Trousseau relates that in the year 1825, two officers quartered at Tours quarrelled, and satisfied their honour by a duel, as a result of which, one of them received a ball which entered at one temple and made its exit at the other. The patient survived six months without any sign of lesion of articulation, nor was there the least hesitation in the expression of his thoughts till the supervention of inflammation of the central substance, which occurred shortly before his death, when it was ascertained that the ball had traversed the two anterior lobes at their centre.

Here then are three cases in which the two anterior lobes, the presumed seat of speech, according to Bouillaud, were both destroyed or very extensively injured. What does a conscientious analysis of them teach us? In M. Peter's case, we have seen that speech was preserved, although both frontal lobes were reduced to a jelly; in Professor Trousseau's case, a ball had traversed the two anterior lobes at their centre, entering at one temple, and making its exit at the other, and speech was also unaffected; whilst in the third case, that of M. Velpeau, although a tumour had actually taken the place of the two anterior lobes, instead of being speechless, the man was remarkably loquacious.

These three cases, to which I could add others, seem to me to upset M. Bouillaud's theory, by showing that a profound lesion may exist in both anterior lobes without impairment of articulate language.*

Having disposed of the theories which locate the faculty of language in one

* I pass over the unilateral theory of Dr. Dax, who places the seat of speech in the left hemisphere, to the exclusion of the right.

or both anterior lobes, I arrive at the consideration of the views of Professor Broca, the perpetual secretary of the Anthropological Society of Paris, whose researches lead him to confine the seat of speech to a very narrow limit, a particular fold of the left anterior lobe, called the *third left frontal convolution*, and which is marked 3 in Figure IX.

Of all the theories that have been advanced, this least of all will stand the test of an impartial scrutiny, and evidence is daily accumulating of such a nature as to undermine M. Broca's position at every point. I have in another work discussed the value of this theory at considerable length;* I will simply state here that I

---

\* "On Aphasia, or Loss of Speech, and the Localisation of the Faculty of Articulate Language," Pp. 155 —160. Churchill and Sons, 1870.

have myself met with cases of loss or impairment of language, in which this particular fold was found quite healthy; furthermore, one case has been observed by M. Moreau, of Tours, in which this convolution was *congenitally absent*, and yet the patient showed no symptom of loss of language. Now, I need not dwell further on this hypothesis, for it must be apparent to everybody that the cases I have quoted of destruction of the anterior lobes apply equally, or I may say *a fortiori*, to this theory; for, what proves the greater proves the lesser, and it is not conceivable that M. Broca's pet fold can have escaped injury amid the general destruction caused by the lesions described.

I cannot dismiss this hypothesis without calling attention to the confirmation that would be given to Mr. Darwin's views if

M. Broca's theory were correct, and this particular fold could be shown to be the seat of speech in man. And here I must call attention to the comparison which Carl Vogt makes between our quadrumanous cousins and ourselves. According to this distinguished naturalist, the apes have an extremely imperfect development of the third frontal convolution, and the same condition exists in the microcephali; therefore, he says, as neither apes nor microcephali can speak, Comparative Anatomy gives a subsidiary support to the theory which places speech in this convolution.

I have been in communication with Professor Vogt in reference to this subject, and he has kindly favoured me with his views, which I consider so extremely pertinent to our subject, that I shall give them in his own words, as contained in an autograph letter to myself:—

'The brain of man and that of apes, especially of the anthropoid apes (orang, chimpanzee, gorilla), are constructed absolutely upon the same type— a type by itself, and which is characterised, amongst other things, by the fissure of Sylvius, and by the manner in which the island of Reil is formed and covered; thus in man, the third frontal convolution is extraordinarily developed, and covers partly the insula, whilst the transverse central convolutions are of much less importance. In the ape, on the other hand, the third frontal convolution is but slightly developed, whilst the central transverse convolutions are very large.

To show the bearing all this has upon the seat of speech, I would refer to the microcephali, who do not speak; they learn to repeat certain words like parrots, but they have no articulate language. Now, the microcephali have the same conformation of the third frontal convolution as apes; they are apes as far as the anterior portion of their brain is concerned. Thus, man speaks; apes and microcephali do not speak. Certain observations have been recorded which seem to place language in the part which is developed in man, and contracted in the microcephali and in the ape; Comparative Anatomy, therefore, comes in aid of M. Broca's doctrine.'

I have reason to believe that these views of Professor Vogt are not very generally known in this country; and I need hardly allude to the extremely important bearing they have upon the question at issue; for if Professor Broca's theory could be proved to be correct,—that this third frontal convolution is the seat of human speech,—a strong argument could be adduced in favour of Darwinism. It might be said that the ape possessed the rudiments of speech in an undeveloped form, and that in subsequent generations, by the process of evolution, this fold would become more developed, and the ape would speak; in fact, would become a man! As, however, this fold has *not* been proved to be the seat of speech in man, the Darwinian argument from analogy of structure falls to the ground, and speech remains a barrier the brute is not destined to pass.

# CHAPTER VI.

Opera naturale è ch' uom favella :
Ma così, o così, natura lascia
Poi fare a voi secondo che v' abbella.
DANTE, *Del Paradiso,* xxvi., 130.

*Language is a Distinctive Attribute of Man— Man versus Ape controversy—On the Universality of Language—Is there a Speechless Tribe?—The Fuegians and the Veddahs of Ceylon — Tylor, Lubbock, Whitney, and Trench — The so-called speechless wild Men were probably Apes—Evidence of the great travellers of the day.*

I think I have now established my first proposition, by showing that in Articulate Language, Man has a faculty not shared by animals; in fact, that Articulate Speech is a Distinctive Attribute of Man, thus establishing a difference of *kind* between him and the brute.

I have gone thus minutely into the question of the Seat of Language, in order to demonstrate its psychological nature, and thus support my third proposition, viz., the Immateriality of the Faculty of Speech, and also to show that no arguments founded on the analogy between the physical structure of the brain of man and that of the ape, can be brought forward against my views. Now this very objection was raised by a gentleman who took a part in a controversy between myself and others, which was carried on during the summer of 1872, in the columns of the *Eastern Daily Press*, under the title of Man *versus* Ape.*

* I have thought it right to review the main features of this controversy, the interest in which was not confined to East Anglia, as shown by the fact that several gentlemen from a distance took part in it.

Early in the spring of 1872, I had the opportunity of submitting my views to the consideration of one of the learned societies of the metropolis, when I endeavoured calmly and dispassionately to explain the reasons which induced me to prefer the Mosaic account of the origin of man to the hypothetical statements of Mr. Darwin and his disciples. On my return home, however, I was at once assailed by a host of adversaries, who, with great warmth, resented my insolent attempt to deprive them of their claim to be allied to the monkey tribe! I was accused of bigotry and superstition, and was regarded as one of the narrow-minded disciples of the "extinguished theologians that lie about the cradle of every science as the strangled snakes beside that of Hercules."

Now, I desire to give the writers of the various communications credit for the great ability and perseverance they have shown in so stoutly defending their claim to a descent from an Anthropoid Ape; the general ardour displayed has only been equalled by those, who, in former days, strove with so much anxiety to trace their ancestry to the roll of Battle Abbey. But, whilst fully recognising the talent, and in one instance at least, the great geological knowledge displayed by these claimants to Simian ancestry, I most unhesitatingly affirm that they have not, in the smallest degree, weakened the position taken by me in my paper at the Victoria Institute,* which was that in

* In order to show that this is not a mere empty boast on my part, I challenge a reference to the numerous letters on this subject which appeared in

language, we possessed a difference of *kind* between man and the ape which Mr. Darwin professes his inability to find.

That portion of the controversy which had reference to the Missing Link, or the absence of any intermediate forms between man and his supposed progenitors, either in a living state or in a fossil condition, was very ably dealt with by the Rev. W. P. Lyon, and is published in the third edition of his book, entitled "Homo *versus* Darwin," a work which I can heartily recommend as containing a clear and logical refutation of the evolution theory, at all events,

the *Eastern Daily Press*, from March 27th to July 13th, 1872; and I beg those who may thus care to review this correspondence, to discard from their minds anything I may have said which does not fully commend itself to their impartial judgment.

in its application to Man. To that portion of the correspondence which has reference to Language, I must just briefly allude.

My first opponent, Mr. A—— enters the list with the assertion that language is not an attribute *universally* belonging to the human race, and that there are tribes of savages who have "*nothing of the kind*," adding, that if such be the case, "Dr. Bateman's argument falls to the ground." Of course it does, and I stake my anti-Darwinian position upon the point thus raised. I have been all along assuming that Articulate Language is a universal attribute of Man, and I need not say that if it can be shown that such is not the case, my statement, that Language constitutes a difference of kind between Man and animals is at once controverted. The problem which I desire to solve is, whether any direct evidence can be found of the existence of races of men, past or present, who are without language of any kind; whether, in fact, speech is universal amongst mankind.

Let us see what evidence Mr. A—— adduces against my second proposition; viz., that articulate speech is a universal attribute of Man. He refers me to a well-known book of travel, the "Voyage in the *Beagle*," where it is stated that the Fuegian savages can only *cluck* like a hen. Now, I have referred to the passage to which my attention is called, and I find that this description of the Fuegian savages is by Mr. Darwin himself, who was the naturalist to the expedition in which the *Beagle* was engaged. From Mr. Darwin's account of this singular race, it is evident that they *did* possess articulate speech, for although they gave no evidence of conversational powers, Mr. Darwin says, " They could repeat with perfect correctness each word in the sentence addressed to them, and they remembered such words for some time." Hence it is evident that they possessed the faculty of language, although in an imperfectly developed form. Now these Fuegians are described in " The Descent of Man," as ranking amongst the lowest barbarians.* Captain

* 'The astonishment which I felt on first seeing a party of Fuegians on a wild and broken shore will never be forgotten by me, for the reflection at once rushed into my mind—such were our ancestors. These men were absolutely naked and bedaubed with paint, their long hair was tangled, their mouths frothed with excitement, and their expression was wild, startled, and distrustful. They possessed hardly any arts, and

Mayne Reid also bears similar testimony as to their degraded condition, for in his "Odd People," at page 476, he says, "fairly examined in all his bearings, fairly judged by his habits and actions, the Fuegian may claim the credit of being the most wretched of our race." The lowest barbarians, therefore, not only possess the power of speech, but are capable of even learning a foreign tongue, for those brought over to England in the *Beagle* are actually described as being able to talk a little English; * in fact, the late Admiral Fitzroy tells us that when three years later they were restored to their native

---

like wild animals lived on what they could catch; they had no government, and were merciless to every one not of their own small tribe.' " The Descent of Man," Vol. ii., P. 404.

* "Narrative of the Surveying Voyages of the *Adventure* and *Beagle*," Vol. ii. Pp. 2, 121, and 189.

land, 'they had acquired enough of our language to talk about common things.'

The acquisition of articulate language is, in a great measure, the result of imitation. Bring a Fuegian to England, and give him time, and he will talk, for he possesses the healthy germs of speech, and has the capacity for evolving a language; put a monkey under training for any number of years, and he will never evince the slightest capacity for the acquisition of language.

In a short reply to this opponent, I pointed out the palpable error as to his statement about the Fuegians; but this gentleman does not seem to be easily convinced, for he returned to the charge and in a subsequent letter alluded to 'the immense amount of evidence we possess which proves that many tribes of savages do exist who do not possess articulate speech;' and he supported this statement by a reference to the Veddahs of Ceylon, described in Tylor's " Early History of Mankind." Now, on referring to page 77 of this interesting book, I find the paragraph which has misled my opponent, who evidently quotes only as far as suits his purpose, for if he had turned over another leaf, at page 78, he would then have found that Mr. Tylor totally denies the accuracy of the statement that the Veddahs have no language, and does this by combating the very paragraph which Mr. A—— has quoted, as will be seen by the following extract:—

'Mr. Mercer seems to have adopted the common view of foreigners about the Veddahs, but it has happened here, as in many other accounts of savage tribes, that closer acquaintance has shown them to have been wrongly accused. Mr. Bailey, who has had good opportunities of studying them, contradicts their supposed deficiency in language, with the remark that he never knew one of them at a loss for words sufficiently intelligible to convey his meaning, not to his fellows only, but to the Singhalese of the neighbourhood, who are all more or less acquainted with the Veddah patois.' Furthermore, I may add that Mr. Tylor has entered into this question of the universality of the faculty of speech in a most exhaustive manner; he has consulted a variety of authors, and being fully impressed with the recklessness with which assertions are made about savage tribes, he evidently places no reliance in those far-fetched travellers tales, which tell us of natives who have no language, or none that can express anything higher than what we might conceive to be expressed by the neighing of a horse, the cackling of a hen, or the grunting of a hog. 'We have,' says Mr. Tylor, 'no evidence of man ever having lived in society without the use of spoken language; but there are some myths of such races, and, moreover, statements have been made by modern writers of eminence as to an intermediate state between gesture-language and word-language, which deserve careful consideration.' *

* The geographer Pomponius Mela says that in Ethiopia there dwell dumb people, and such as use gestures instead of language (muti populi, et quibus pro eloquio nutus est). Pliny, also, describes tribes who have for their language nods and gestures (quibusdam pro sermone nutus motusque membrorum est).

In another work the same author says that 'popular opinion has under-rated the man as much as it has over-rated the monkey. We know how sailors and emigrants can look on savages as senseless ape-like tribes, and how some writers on anthropology have contrived to make out of the moderate intellectual difference between an Englishman and a negro something equivalent to the immense interval between a negro and a gorilla. Thus we can have no difficulty in understanding how savages may seem mere apes to the eyes of men who hunt them like wild beasts in the forest, who can only hear in their language a sort of

Tylor, in commenting upon these statements, says that 'to go thoroughly into the discussion of these stories would require an investigation of the whole subject of the legends of monstrous tribes.'—*Early History of Mankind*, P. 76.

irrational gurgling and barking, and who fail totally to appreciate the real culture which better acquaintance always shows among the indigenous tribes of man.' *
From the above passages, it will be seen that Mr. Tylor has arrived at the very opposite conclusion to that imputed to him by Mr. A——.

Another British ethnologist, Sir John Lubbock, speaks in a no less decided tone in reference to the point in dispute. 'Although,' says he, 'it has been at various times stated that certain savage tribes are entirely without language, none of these accounts appear to be well authenticated, and they are *a priori* extremely improbable. At any rate, even the lowest races of which we have any satisfactory account possess a

* "Primitive Culture," Vol. i. P. 342.

language, imperfect though it may be, and eked out to a great extent by signs. I do not suppose, however, that this custom has arisen from the absence of words to represent their ideas, but rather because in all countries inhabited by savages the number of languages is very great, and hence there is a great advantage in being able to communicate by signs.' *

The great American authority in linguistic science, Professor Whitney, thus writes : — 'Language, articulate speech, is a universal and exclusive characteristic of man : no tribe of human kind, however low, ignorant, and brutish, fails to speak ; no race of the lower animals, however highly endowed, is able to speak : clearly,

* " The Origin of Civilisation and the Primitive Condition of Man," P. 313.

it was just as much a part of the Creator's plan that we should talk as that we should breathe, should walk, should eat and drink. The only question is, whether we began to talk in the same manner as we began to breathe, as our blood began to circulate, by a process in which our own will had no part; or, as we move, eat, clothe and shelter ourselves, by the conscious exertion of our natural powers, by using our divinely-given faculties for the satisfaction of our divinely-implanted necessities.' *

Archbishop Trench, in refuting the notion that language was invented by man himself, and that it must therefore be put on the same level with the various arts and inventions with which man has gradually adorned and enriched his life,

* "Language and the Study of Language," P. 399.

goes on to say that 'language would then be an *accident* of human nature; and, if such were the case we certainly should find tribes sunken so low as not to possess it, even as there is no human art or invention, though it be as simple and obvious as the preparing of food by fire, but there are those who have fallen below its exercise. But with language it is otherwise. There have never yet been found human beings, not the most degraded horde of South African bushmen, or Papuan cannibals, who did not employ this means of intercourse with one another.'\*

I am quite aware that books of travel abound with tales of wild men without the use of speech — men who whistle like birds and shriek like apes; the

\* " On the Study of Words," P. 12.

disputant in the 'Man *versus* Ape' correspondence, to which I have before alluded, has called my attention to this fact, but a careful scrutiny of these statements will at once demonstrate their untrustworthy character. It is evident that the so-called speechless wild men of certain authors were in reality apes of some large species. Dr. Livingstone, in his interesting account of the region of the Manyuema, describes an ape-like creature called Soko whom the natives regard as belonging to the human species; some of them believing that their buried dead rise as Sokos. This animal often goes erect but places its hand on its head as if to steady the body. In speaking of this creature, the natives are in the habit of saying ' Soko is a man, and nothing bad in him.' From Livingstone's description, however, it is clear that the Soko is merely a new species of anthropoid ape.*

Herodotus, speaking of a tract of land in Lybia abounding with forests and wild beasts, describes a race of wild men and wild women, but there can be but little doubt that these creatures were apes of some large species. Rawlinson evidently adopts this view, as shown by his contrasting the description of Lybia as given by Herodotus with that contained in the voyage of Hanno:—'At the bottom of this bay lay an island, having a lake, and in this lake another island, full of wild people. (μεστὴ ἀνθρώπων ἀγρίων). Far the greater proportion were women whose

---

\* " Livingstone's Last Journals," Vol. ii. P. 52.

bodies were covered with hair and whom our interpreters called Gorillæ.'

Lord Monboddo says, ' not only solitary savages but a whole nation, if I may call them so, have been found without the use of speech.' He, however, deprives this statement of any force it might otherwise have by the next paragraph in which he says, ' This is the case of the Orang Outangs that are found in the kingdom of Angola in Africa, and in several parts of Asia. They are exactly of the human form; walking erect, not upon all-four, like the savages that have been found in Europe. I was further told, by a gentleman who had been in Angola, that there were some of them seven feet high, and that the negroes were extremely afraid of them; for, when they did any mischief to the Orang

Outangs, they were sure to be heartily cudgelled when they were caught. But though, from the particulars above-mentioned, it appears certain that they are of our species, and though they have made some progress in the arts of life, they have not advanced so far as to invent a language; and accordingly none of them that have been brought to Europe could speak, and, what seems strange, never learned to speak.' Monboddo labours hard to establish a relationship between the Orang and Man, and has a long chapter about this homo tetrapus, mutus, hirsutus, which, according to him, is a 'barbarous nation which has not yet learned the use of speech.'*

A few years since, the Rev. Dunbar

* 'The Origin and Progress of Language,' Vol. i. Pp. 187, 270.

Heath read a paper before the Anthropological Society of London, in which he mooted the hypothesis that the original inhabitants of Europe, the contemporaries of the woolly elephants and of the rhinoceroses were mutes, and he suggested that man may have existed over vast areas and during long periods in this 'mute emotional state,' and that traditional notions constitute the only reason why this idea should not be accepted.*
He then goes on to say that 'man's triumph, language, is generally supposed to have happened simultaneously with another great event, namely, the very

* Horace, in one of his satires, speaks of men as having been originally speechless — mutum et turpe pecus—and then he goes on to describe them as fighting amongst themselves for shelter and sustenance until they invented a language:—

Donec verba, quibus voces sensusque notarent,
Nominaque invenere.

first appearance in the kosmos of the being now called man, with bigger brain, shorter arm, and stouter thigh than a set of other beings called apes, who had long been domiciled in the neighbourhood of Paradise.'* In another communication to the same society, Mr. Heath supposes that these apes were the ancestors of European men, who were at first dumb, but who in the course of time gasped for articulation and obtained it. 'I confine myself,' says Mr. Heath, 'to the accepting and explaining known and knowable phenomena. It is known that anthropoids existed throughout Europe. It is knowable that they became mute men. It is knowable that these mutes gasped

* "Journal of the Anthropological Society of London," Vol. v, P. 83.

after articulation, and in a few spots attained to it.' *

In the discussion that followed these remarkable communications, one speaker suggested that the Society should establish a colony of Orang Outangs, and that Mr. Heath should teach them to speak. I cannot learn, however, that this practical suggestion of his brother anthropologist has been carried out by Mr. Heath.†

\* " On the Primary Anthropoid and Secondary Mute Origin of the European races." — *Anthropological Journal*, Vol. iv. P. 33.

† At the same debate, a gentleman resuscitated Dr. Adam Clarke's theory that the larger apes had once the gift of speech, and that the reason they do not speak now was, that it was an Orang Outang and not a serpent that tempted Eve, and that the gift of speech was therefore taken from the apes as a punishment. Another speaker, however, pointed out that the facts of natural history were directly opposed to this view, as anthropoid apes do not exist in Arabia nor in Persia, but exist naturally only in tropical regions.

M

The question as to whether language is an attribute universally possessed by the human race is such an important one, as far as the present controversy is concerned, that I wished to corroborate my views by an appeal to the great travellers of the day.

My first correspondent was Dr. Moffatt, the distinguished African traveller, whose long residence amongst savage tribes renders his testimony peculiarly valuable, and his opinion is very decided in reference to the particular point I am now discussing, as will be seen by the following letter with which he has kindly favoured me.

Brixton, June 13th, 1872.

Dear Dr. Bateman,

The Darwinian theory is altogether so ludicrous that I never can refer to it from the platform, which I sometimes do, without taxing the risible powers of my audience. I have had a great deal to do with apes and especially with baboons, and once had to dispute with a company of them the right to a drink of water, but nothing was heard on their part but the everlasting grunt. Mr. Darwin might have selected some more sensible brute to establish his hypothesis, for the mind is the standard of the man.

With regard to speech being the dividing point between man and the brute, I perfectly agree with you. This barrier has never been, nor ever can be, overleaped, and it appears to me extraordinary that any one can think otherwise. I have had much intercourse with the bushmen in the interior of South Africa, and they may be set down as the lowest grade of humanity in that country. In some respects, their language has a resemblance to the clicking language of the Hottentots. It is much more guttural, and enunciated a good deal through the throat, and not understood by the Hottentot. Even among themselves, the bushmen of one district do not understand those of another living at no great distance. I have frequently listened to their conversations, when there appeared to be no difficulty whatever in communicating their ideas to each other. When taken into service, they readily learn to speak fluently the languages of English, Dutch, and Sechuana. They are certainly the most degraded race to be found in the interior. Villages, folds, or flocks, they have none, but move about in search of game, roots, wild honey, and are emphatically children of the desert.

Of all the reports I ever heard respecting interior tribes, I never found that the idea was ever entertained that human beings existed that did not possess a language, and ability to convey their ideas with perfect clearness.

By-and-by, when Dr. Livingstone shall arrive among us, he will no doubt tell us strange things; but nothing, I believe, that can possibly sanction Darwinism.

I am, my dear Sir, yours, &c.,

ROBERT MOFFATT.

The testimony of Sir Bartle Frere is equally unequivocal, as will be seen by the following communication which he has courteously addressed to me:—

22, Princes Gardens,
4th December, 1873.

MY DEAR SIR,

I have just received yours of the 3rd of December, and hasten to assure you that I believe you are perfectly right in your conclusions. The tribes to which you refer are to be found in almost all the very dense forest parts of India under different names, and apparently of different origin. In the jungles of Eastern Bengal and Central India, and also in some of the dense forests on the skirts of the Nilgherry Hills and in Ceylon are, or were within the last fifty years, forest tribes who wear little or no clothing, and live in trees, but all have a language of their own which, however imperfect for expressing any ideas beyond those of savage life, is quite sufficient for their purpose, and entitles them to be included in the species of "articulate-speaking men," one of the descriptions which, I believe, as you rightly suppose, correctly defines the limits between man and beast.

Truly yours,
H. B. W. FRERE.

Sir Samuel Baker, than whom no one is better qualified to speak authoritatively on this subject, writes to me as follows:—

> Almond Hotel, Clifford Street, W.,
> December 20th, 1873.
>
> DEAR SIR,
> I have never heard of a speechless tribe; nor do I believe such savages exist. All those I have actually visited not only have speech, but also numerals. They usually count in tens, taking for the base of their calculations their digits, which appear to be the original root of numbers.
>
> Very truly yours,
> SAMUEL W. BAKER.

From the summary which I have thus endeavoured to give of the researches of the most trustworthy of ancient and modern writers, and from the evidence furnished by Tylor and Lubbock, who may be considered as representative men in Ethnology, supported as it is by the testimony of the great travellers of the day, it will be seen that all militates against the notion of the existence of a speechless tribe, and confirms the truth of my second proposition, which is that Articulate Speech is a Universal Attribute of Man, and that the wildest savage that roams the woods in still undiscovered lands, has a language or the capacity for acquiring it.

# CHAPTER VII.

> "We may analyse the sun and penetrate the stars, but man is conscious that he is made in God's own image."
> EARL OF BEACONSFIELD.—*Lothair*.

*The Immateriality of the Faculty of Speech—The Brain a mere Instrument—The Electric Telegraph and its Language—Inconsistencies of the Evolutionists—The Odium Antitheologicum—The Mystery of Life—Conclusion.*

The main object of this treatise has been to test Darwinism by Language— to examine the Evolution theory from a linguistic point of view, and to see whether the attribute of Articulate Speech establishes a difference of *kind* between man and animals.

My first point has been to show, and I must leave it to my readers to judge how far I have succeeded in showing, that animals do not possess a trace of articulate language, and that Speech is a Distinctive Attribute of Man; if this be so, the faculty of language establishes a difference between man and animals, not of *degree* only, but of *kind*, in fact, the very difference which Mr. Darwin has been so long in search of, and which he has hitherto failed to discover.

The enunciation of my first postulate would have influenced the question I am discussing but very little, unless I could also establish my second proposition, viz., that Articulate Speech is a Universal Attribute of Man. I have entered into this feature of the controversy at considerable length, and I have conclusively shown that no reliable evidence can be adduced of a Speechless Tribe.

For the purpose of establishing my third proposition, viz., the Immateriality of the Faculty of Speech, it has been necessary to enter briefly into the much-vexed question of the Seat of Speech — the Localisation of the Faculty of Articulate Language; for, as the remarkable similarity between the brain of man and that of the ape cannot be disputed, if the seat of human speech could be positively traced to any particular part of the brain, the Darwinian could say that although the ape could not speak, he possessed the *germ* of that faculty, and that in subsequent generations, by the process of evolution, the "*speech centre*" would become more developed, and the ape would then speak.

I have endeavoured, however imperfectly, to show that none of the various theories as to the seat of language will stand the test of an impartial scrutiny. I have shown, and that upon the most indisputable authority, that persons could talk when the *presumed* seat of speech was invaded by an enormous tumour, completely disorganised by disease, or destroyed by a pistol-shot!

With these facts before me, I am tempted to ask whether speech, like the soul, may not be an attribute—an immaterial *nescio quid*, the comprehension of which is beyond the limits of our finite minds?

When we talk about the faculty of speech, have we any clear and definite notions as to what we mean? Does the loss of it necessarily imply organic lesion of structure — *material damage?* * If it were so, how can we account for the cases recorded in which the restoration of the power of speech was due to the effect of a severe mental shock?

We are all familiar with the story in Herodotus of the son of Crœsus, who had never been known to speak, but who, at the siege of Sardis, being overcome with astonishment and terror at seeing the king, his father, in danger of being killed by a Persian soldier, exclaimed aloud,

---

* In those cases of loss of the power of speech where there is no evidence of organic lesion, the defect may possibly be due to some chemical, thermal, or electrical change in the brain tissue. To discuss this interesting point would be to transgress the proper limits of this treatise, and the author must refer his readers to his work on "Aphasia" for further information as to the cause of impairment of Language in those cases, where there is no altered state of the cerebral structures, appreciable to the sense of vision.

"Ἄνθρωπε, μὴ κτεῖνε Κροῖσον—Oh man! do not kill Crœsus. This was the first time he had ever articulated, but he retained the faculty of speech from this event as long as he lived. Herodotus is universally admitted to be a trustworthy historian; but if it be thought far-fetched to illustrate a subject by allusion to a work written 500 years before the Christian era, I may add that such cases have been met with by modern observers. My friend, Mr. Robert Dunn, has recorded a similar one, and I myself was recently requested to see a man who had suddenly become speechless; the suspension of the power of speech was unaccompanied by any symptom of paralysis, and the loss of the faculty of articulate language continued for six days, when, being asleep on his couch, he suddenly started up, and was heard to say three times, "A man in the river!" From this moment speech was restored, and when I saw him an hour afterwards, he told me that he had dreamed that a man was falling into the river. The mental shock produced by this dream was salutary, for it resuscitated the previously dormant faculty of articulate language.

Surely we cannot, for one moment, assume that in these cases there can have been any structural lesion of the brain, any *material* damage.

But I may be told,—granted the truth of your statements, surely you must admit that man speaks by and through his brain. Most assuredly I do. Man in this life thinks and wills by means of his brain, which is undoubtedly the material organ of the mind, or, to use the language of one of our veteran psychologists, 'the vesicular matter of the encephalic ganglia is the *material substratum* through which all psychical phenomena of whatever kind, and among all races of mankind, are manifested in this life.'

Every faculty manifests itself by means of matter, and the material condition which renders the exercise of a faculty possible is an *organ*, and it is important not to confound the faculty with the corporeal organ upon which the external manifestation of this faculty depends. The muscles of the body are the means by which we exercise the power of motion, but it would be illogical to say that the muscles were the seat of *the vital force* by which we move about. Again, by means of the Electric Telegraph, ideas and words are transmitted from mind to mind with a rapidity to which human speech cannot attain. Now the electrical battery may be not inaptly compared to the brain, and the telegraph wires to the nerves which emanate from the cerebral organ to supply the various structures engaged in articulation. If the battery is out of order, or the telegraphic wires are broken, this "lightning language" by which mind speaks to mind, becomes impossible. Precisely in the same way, a certain normal and healthy state of cerebral tissue is necessary for the exterior manifestation of the faculty of speech, but that is a very different thing from saying that speech is located in this or that particular portion of the brain, or that Language is but the corresponding result of a certain definite molecular condition of the cerebral organ.*

Perhaps I cannot better illustrate my meaning than by an allusion to a passage in Plato's celebrated dialogue on the Immortality of the Soul, where a disputant with Socrates inquires if the soul is not like the harmony of a lyre, more beautiful, more divine than the lyre itself, but yet

---

* 'No man of any philosophic culture, says Max Müller, ' will look on the brain, or that portion of the brain which interferes with rational language, as the seat of the faculty of speech, as little as we place the faculty of seeing in the eye, or the faculty of hearing in the ear. That without which anything is impossible is not necessarily that by which it is possible. We cannot see without the eye, nor hear without the ear, but neither can the eye see without us, or the ear hear without us. To look for the faculty of speech in the brain would, in fact, be hardly less Homeric than to look for the soul in the midriff.' "Lectures on Mr. Darwin's Philosophy of Language," *Fraser's Magazine*, Vol. vii., P. 676.

is nothing without the lyre, vanishing when this instrument is broken. For the word *soul*, substitute *speech*, and for *lyre*, substitute *brain*. The iustrument, *i.e.* the brain, may be damaged, and speech may become impossible, but that does not constitute the brain the *seat* of speech, although it is undoubtedly the *instrument* by which this attribute becomes externally manifested.

Although my chief aim has been to examine the Darwinian theory from a linguistic point of view, it will be seen that I have been tempted to digress somewhat from my original intention, and to consider the general subject of Evolution in all its bearings.

I. now desire briefly to point out what seem to me to be certain inconsistent and illogical features in the position assumed by some of the members of this modern school of thought.

Whilst wishing to handle this controverted subject in a spirit of fairness and impartiality, I must enter my protest against the extremely illiberal attitude assumed by some of the Evolutionist writers — an attitude which savours of sharp and clever diplomacy, rather than of fair and honourable discussion.

One striking characteristic of their tactics is the confidence and admiration they express towards all who agree with them, their writings being stamped with the most fulsome eulogy of each other, and with gross abuse of their opponents, together with unseemly discourtesy towards all those who venture to differ from them;* in fact, they seem, in many instances, to lay themselves open to the charge of seeking victory in argument rather than the triumph of truth; and especially do they evince the most profound eagerness to discover, if possible, some vantage ground for an attack on religious belief. 'It is easy,' says Mivart, 'to complain of the one-sidedness of many of those who oppose Darwinism in the interest of orthodoxy; but not at all less patent is the intolerance and narrow-mindedness of some of those who advocate it, avowedly or covertly, in the interest

* This feature of the controversy is well portrayed in an article on "Modern Scientific Materialism," in *Blackwood's Magazine* for November, 1874, in which the writer calls attention to the fact that 'names, however unknown, if only associated with some attack on theology, or some advance of materialistic speculation, are brought into the full blaze of applausive recognition.'

of heterodoxy. If the *odium theologicum* has inspired some of its opponents, it is undeniable that the *odium antitheologicum* has possessed not a few of its supporters.'\*

This antagonism to religion is patent in the writings of many of the evolutionist school, who appear more anxious to undermine religious belief, than to resolve scientific problems; and although they are constantly accusing theologians of illiberality, they seem themselves to write, as it were, under the yoke of a preconceived opinion; they ransack the storehouse of natural science for weapons against Holy Writ, they unfurl the flag and blow the trumpet of defiance—their motto being *Ecclesia delenda est*.

Professor Haeckel's bias is very apparent, for after drawing a distinction

\* "On The Genesis of Species," Pp. 12, 14.

between what he calls scientific and moral materialism (naturwissenschaftlicher und sittlicher Materialismus), he indulges in the following coarse and most uncalled for tirade against the clergy, and against all forms of religion. 'Moral Materialism,' says he, ' has for its sole object a refined sensual enjoyment. You will seek for it in vain amongst naturalists and philosophers, whose supreme delight is the intellectual contemplation of nature, and whose highest aim is the knowledge of nature's laws. If you wish to find it, you must seek for it in the palaces of princely churchmen, and amongst those hypocrites, who, under the mask of an austere piety, aim only at the exercise of a hierarchical tyranny over their fellow-creatures. Too dull to understand the infinite nobility of what is called 'crude matter,' and to appreciate the glorious phenomena arising out of it, insensible to the inexhaustible charms of nature and ignorant of her laws, they fulminate their anathemas against the whole of the natural sciences, whilst they themselves plunge into the most repulsive form of materialism. It is not only the infallible papacy with its endless chain of horrible crimes, but the perverse moral history of the orthodox in all forms of religion can be cited in proof of what is here stated.' *

A Trans-Atlantic author writes in a no less illiberal and petulant strain, 'Religion,' says he, 'must relinquish that domineering position which she has so long maintained against Science. The ecclesiastic must learn to keep himself

* " Natürliche Schöpfungsgeschichte," P. 33.

within the domain he has chosen, and cease to tyrannise over the philosopher, who, conscious of his own strength, and of the purity of his motives, will bear such interference no longer.*

I need scarcely add that the cause of truth is not likely to be advanced by such rhetorical farrago, or rather, I should say, by such coarse and vulgar abuse, as that contained in the above extracts.

It has always appeared to me to be a most strange and inexplicable peculiarity on the part of certain writers of the modern school of thought, that they systematically deprecate any attempt to reconcile Science and Scripture. They willingly concede to the free-thinkers of

---

* "The conflict between Religion and Science," by J. W. Draper, M.D., LL.D., Professor in the University of New York, 1876, P. 367.

the day, the right to use Science for the purpose of subverting religion, but they look with a jealous eye upon those who seek to point out the analogy between the two. May I ask them what value they would attach to any work on the early history of our island, that contained no allusion to Cæsar's Commentaries; and, surely, it would be equally monstrous to consider any theory as to the origin of Man, without, at least, a reference to the Book of Genesis—the first, if not the only book, which professes to enlighten the human race as to its origin.

I, myself, have been accused of using Scripture to refute Darwinism. I beg to say I do nothing of the kind, and there is nothing in this essay to justify such a construction. I use Science to show that language is the difference of *kind* between man and animals, which Mr. Darwin seems to stand in need of; and having, however imperfectly, combated his views from a linguistic point of view, I *incidentally* call attention to the fact that Science corroborates Holy Writ, just as Bishop Colenso and others contend that it controverts it. This is a very different thing from the illogical process imputed to me of bolstering up scientific views by appealing to the authority of Scripture, which I freely admit was never intended to teach us Science.

I doubt not that many of those who have differed from me are serious, thoughtful men, who would not knowingly propagate a dangerous doctrine; but I must think they cannot have realised the ultimate consequences of their proposal to ignore the Book of Genesis in any search after truth, simply because, in such a search, the aid of Science may also be required.

No person regrets more than I do the tendency of the present day to throw Theology and Science into two opposite and contending parties, but, surely, no scientific deduction is of less force or value because it is shown to be in harmony with Revelation, and the remarks of Bishop Temple are just as applicable to the scientist as to the theologian, when he says 'He is guilty of high treason against the faith, who fears the result of any investigation, whether philosophical, or scientific, or historical.'

There is another class of reasoners who assume an attitude of indifference in regard to this subject, urging that the great truths of Scripture cannot be seriously affected by the evolution theory, since many sound theologians no longer contend for the literal and verbal inspiration of the Bible. Now, this is not a question of mere *verbal* accuracy.* Darwinism is not merely inconsistent with this or that particular line or passage, but is incompatible with the whole spirit of the Bible, where at almost every page, the idea of a personal Creator is implied; whereas the evolution theory abolishes all idea of creation in the ordinary sense of the term. The aim, end, and ultimate consequences of this doctrine are well set forth in an article in the Transactions of the Victoria Institute, where the author thus describes what he calls the scientific creed of modern

* In support of the above view, see Lord Hatherley's work on "The Continuity of Scripture," also Archdeacon Pratt's "Scripture and Science not at Variance."

Anthropology :—" I believe in Law, but no Lawgiver; in the life-giving power of Force and Substance; Intelligence from Non-Intelligence, without conscious Author. I believe in the natural cohesive magnetic formation of the earth on which I dwell, and the origin of Man from Beast; the never-ending development of species in animated nature generally, first by Spontaneous Generation, afterwards Natural Selection. I believe in the eternity of matter, which sets itself in motion, and governs all worlds, and I look for the oldest Homo Sapiens in pliocene or miocene strata, and that his fossilised bones will be found, on examination, to be either those of an Ape more anthropoid, or a man more pithecoid, than any yet known, Neanderthal or Engis cranium notwithstanding. I also believe in the sure mortality of the Human Soul, which is but an attribute of Brain-Protoplasm." *

In all that has been said and written about Evolution, I have been struck with the complete absence of facts—everything is hypothetical. The evolutionists deal largely in the subjunctive mood,—the *may* and the *might*—and on purely hypothetical premises, they attempt to found conclusive arguments. If we strip their assertions of all their vagueness and superficial varnish, and reduce them to a skeleton of logical statement, we shall see how much is assumed and how little proved, and we shall also find that we are asked to accept a chain of hypotheses, as if it were an

---

\* "Journal of the Transactions of The Victoria Institute, or Philosophical Society of Great Britain," Vol. v. P. 265.

induction founded on ascertained and indisputable facts.

As a remarkable instance of the above fallacious mode of reasoning, I would cite Strauss, who, in a chapter devoted to the consideration of the doctrine of Evolution and of the manner in which the universe has been formed, whilst quoting Virchow to the effect that spontaneous generation does not *now* take place, very speciously insinuates that it *may* have occurred in some other epoch of the world's history, and that we have no evidence that it did *not* occur in some primeval period, when the world was in a totally dissimilar condition.* And upon this unwarranted surmise,—on this monstrous guess, he builds a Universe, and all that in it is!

Mr. Darwin himself does not pretend to

* "Der alte und der neue Glaube," Bonn, 1873, P. 174.

prove anything, all that he claims for his theory is that it is possible, but his disciples declining to accept the *onus probandi*, maintain that Mr. Darwin's explanation ought to be accepted as true, unless some more plausible theory be advanced, but, as one of his critics justly remarks, ' surely, this is to mistake altogether the object of scientific inquiry, for it by no means follows that an improbable hypothesis ought to be accepted, because its opponents are unable or unwilling to propose a new hypothesis several degrees less improbable.' *

Again, some writers imply that Evolution must be true because certain scientific celebrities believe in it, thus setting aside the right of private judgment, and claiming dominion over our faith, on the

* *British Quarterly Review,* October, 1871, P. 464.

authority of men of high scientific attainments — men, however, who view everything through a biological medium. Their great *cheval de bataille*, is Sir Charles Lyell, and they are for ever reminding us that although in all the early editions of his 'Principles of Geology,' he looked upon geological facts as proving the fixity of species and their special creation in time, yet in the 10th edition, he announces his change of opinion, and his conversion to the doctrine of development by law. Now, in thus dwelling with such complacency on the so-called conversion of the Nestor of geologists, the evolutionists fall into the too common error of confounding facts themselves with deductions drawn from these facts, for as an American writer, Professor Hodge, very justly remarks 'the change on the part of this eminent geologist, was a mere change of opinion; there was no change of the facts of geology between the publication of the eighth and of the tenth edition of his work, neither was there any change in his knowledge of those facts. All the facts relied upon by evolutionists have long been familiar to scientific men. The whole change is a subjective one. One year the veteran geologist thinks the facts teach one thing, another year he thinks they teach another. It is now the fact, and it is feared it will continue to be a fact, that scientific men give the name of science to their explanations as well as to the facts. Nay, they are often more zealous for their explanations than they are for the facts.' *

* 'What is Darwinism'? by C. Hodge, D.D., LL.D., P. 134.

I do not wish to imitate the example of some of my opponents, by making this a question of arithmetic; I may say, however, that I by no means agree to the statement that scientific men generally are in favour of Evolution, as a large number of the foremost naturalists and physiologists of the day, many of whose writings I have quoted in this essay, are utterly opposed to it.

The late Professor Agassiz, usually described as the Cuvier of America, thus writes ' Were the transmutation theory true, the geological record should exhibit an uninterrupted succession of types blending gradually into one another. The fact is that throughout all geological times, each period is characterised by definite specific types, referable to definite orders, constituting definite classes and definite branches built on definite plans. Until the facts of nature are shown to have been mistaken by those who have collected them, and that they have a different meaning from that now generally assigned to them, I shall consider the transmutation theory as a scientific mistake, untrue in its facts, unscientific in its method, and mischievous in its tendency.* The same writer, in what I believe was the last production of his pen, says 'As a Palæontologist, I have from the beginning stood aloof from this new theory of transmutation, now so widely admitted in the scientific world. Its doctrines, in fact, contradict what the animal forms buried in the rocky strata of

---

* *The American Journal*, July, 1860, P. 154.

our earth tell us of their own introduction and succession upon the surface of the globe.'*

Principal Dawson, who I am informed is considered as one of the first palæontologists and geologists in America, says 'the evolution theory is itself one of the strangest phenomena of humanity. It existed, and most naturally, in the oldest philosophy, in connection with the crudest attempts of the human mind to grasp the system of nature; but that in our day, a system destitute of any shadow of proof, and supported merely by vague analogies and figures of speech, and by the arbitrary and artificial coherence of its own parts, should be accepted as philosophy, and should find able adherents to string on its thread of hypotheses our vast and mighty

* *The Atlantic Monthly*, January, 1874.

stores of knowledge, is surpassingly strange.' *

I desire to point out what seems to me to be a most illogical feature in the character of a certain school of modern philosophers. They affect to believe nothing, and to be influenced by nothing but what they can fully understand, ignoring the fact that there are certain things which from their very nature are beyond the pale of precise knowledge, and which lie outside the sphere of man's intellect. They take no cognizance of the fact that man is endowed with a *spiritual* nature or moral faculty, wholly independent of the *material* life which he has in common with the rest of creation. Had I not already considerably

* "The Story of Earth and Man," by J. W. Dawson, of McGill College, Montreal.

transgressed the original limits of this essay, I should have liked to have considered the doctrine of trichotomy, and to have discussed the question of the tripartite nature of man—of his possession of the σῶμα, the ψυχή, and the πνεῦμα or organ of God-consciousness, which last differentiates him from the brute which only possesses the σῶμα and the ψυχή. For much valuable information on this important subject I would refer the reader to Mr. Heard's Treatise on The Tripartite Nature of Man, also to some interesting remarks by Sir Tilson Marsh in the Transactions of the Victoria Institute, Vol. V., P. 287.

There is another class of reasoners, who soaring higher into the sphere of 'transcendental obscurantism,' affect the scepticism of the Pyrrhonist school, who maintained that there was no criterion in truth, and whose formula was " We assert nothing—no, not even that we assert nothing." But although they are sceptical upon every other point, they have no doubts whatever about the origin of matter and the genesis of species. Evolution, they cry, magical word, gives us the key to all the mysteries that surround us, enabling us to stride the so-called gulf between mind and matter, and to sweep away the intellectual cobwebs woven by men who lived before the age of enlightenment. Natural Science now teaches us that the difference between so-called organic and in-organic nature is altogether arbitrary, and vital force, as commonly conceived, is a chimera.* There is no distinction between living and dead

* Du Bois Reymond.

matter, and vitality is a metaphysical ghost, (ein metaphysisches Gespenst).

'Life,' says Virchow, 'is only a special, and the most complicated act of mechanics; a portion of the sum-total of matter emerges from time to time out of the usual course of its movements, enters into special organico-chemical combinations, and after having continued therein for a certain time, again reverts to the general modes of motion.'* The brain produces thought just as the liver secretes bile, or as oxygen and sulphur

---

\* 'Das Leben ist nur eine besondre, und zwar die complicirteste Act der Mechanik; ein Theil der Gesammtmaterie tritt von Zeit zu Zeit aus dem gewöhnlichen Gange ihrer Bewegungen heraus in besondre organisch—chemische Verbindungen, und nachdem er eine Zeit lang darin verharrt hat, kehrt er wieder zu den allgemeinen Bewegungsverhältnissen zurück.' *Gesammte Abhandlungen zu wissenschaftlicher Medicin* s. 25. Von R. Virchow.

produce sulphuric acid; in fact, all the varied phenomena of nature are nothing but the molecular changes of matter, and volition and consciousness are mere physical manifestations; give us matter and motion and we will make a Universe!

'If,' says Haeckel, 'anybody feels the necessity of representing the origin of matter as the work of a supernatural creative force independent of matter itself, I would remind him that this idea of an immaterial force creating matter in the first instance, is an article of *faith* which has nothing to do with human science. Where Faith begins, Science ends.' (Wo der Glaube anfängt, hört die Wissenschaft auf.)*

In the above extravagant passage, Professor Haeckel is not consistent with

* "Natürliche Schöpfungsgeschichte," s. 8.

himself, for in the last page of his History of the Creation, he repeats the cry of the philosopher of antiquity, Γνῶθι σεαυτόν, Know thyself. Now let me ask of Professor Haechel, *does he know himself?* Can he understand the mysteries of his own existence, and yet he knows and feels that he lives, although he may not get beyond the formula of Descartes when he said, " Cogito, ergo sum." Can he say that his own existence is merely ' *the product of poetic imagination,*' for that is his definition of Faith ? His text-books of physiology will explain to him all that science can tell him about ontogeny, or the process by which the young of living bodies are produced and their species continued — how the young owe their origin to the evolution of a complex organised structure termed an egg, and how from this egg, under the influence of certain favourable circumstances, the young animal is produced, by an intricate process of vital growth; when all this is learnt, there still remains the Mystery of Life, and man in his perplexity may well say with Coleridge :—

'What is there in thee, Man, that can be known?
Dark fluxion, all unfixable by thought,
A phantom dim of past and future wrought,
  Vain sister of the worm—life, death, soul, clod,
  Ignore thyself, and strive to know thy God!'

Take, again, the vegetable world; a seed which has been for three thousand years buried in the tomb of an Egyptian mummy, is suddenly extricated from its charnel-house, exposed to the influence of atmospheric air and other favourable circumstances, and in due course it becomes a living plant. Now all that science can tell us about this is, that under certain altered physical conditions, the seed has been able to '*germinate.*' Now, what is it that enables the *seed* to germinate, whilst the *stone* remains inactive? What, in short, is the Mystery of Life?

Unlike the philosophers of the present day, the great Sir Isaac Newton, on being asked a similar question, as to why an apple fell to the ground—a fact upon which he founded his grand discovery of the law of gravitation—he replied, 'It is beyond the limit of human reason, it is the will of God.'

One of the most distinguished physiologists of the day, Dr. Beale, in writing upon this subject says, 'there is a mystery in life—a mystery which has never been fathomed, and which appears greater the more deeply the phenomena of life are studied and contemplated. In living centres—far more central than the centre is seen by the highest magnifying powers — in centres of living matter, where the eye cannot penetrate but towards which the understanding may tend—proceed changes of the nature of which the most advanced physicists and chemists fail to afford us the faintest conception, nor is there the slightest reason to think that the nature of these changes will ever be ascertained by physical investigation; inasmuch as they are certainly of an order or nature totally distinct from that to which any other phenomenon known to us can be relegated.'*

In their attempts to gauge the depths of the Universe and to solve the various problems by which they are surrounded,

\* " The Mystery of Life," P. 55, 1871.

philosophers have groped with the taper of science into the dark caverns from whence seem to issue the springs of humanity, but they have failed to explain the Mystery of Life—a theme essentially beyond the grasp of human intellect, and which will not be understood by the loftiest mind in far distant ages, when the scientists of the present day 'like streaks of morning cloud, shall have melted into the infinite azure of the past.'

The question of the origin of the human race has been treated too much as a zoological subject, ignoring the testimony of history, of language, and of other branches of knowledge; Haeckel even forbids the right to speak on this topic to all who are not thoroughly versed in Biology, which he makes the final court of appeal in all scientific matters.

The nineteenth century seems disposed to stake all its hopes on Natural Science, heedless of the fact that Science is ever varying, and that the science of one age becomes the nonsense of the next. I need scarcely add that the Transmutation theory itself is nothing new, for, under the name of metempsychosis, it was in vogue in the earliest times. It is well known that the Egyptians believed that the soul, on leaving the body, passed into the form of some animal, afterwards through the forms of birds and fishes, till it again entered a human frame.

Plato, in his Timæus, makes animals derive their origin from man by successive degradations, the first transition being from man into woman, women being considered as degenerate and effeminate men! The race of quadrupeds sprang from men who had no philosophy, and as they never looked up to the heavens nor cared for celestial objects, their anterior limbs became dragged down to the earth by the force of affinity, and as a necessary consequence of their tastes and occupations. The race of birds was created out of innocent, light-minded men, whose hair became transmuted into feathers and wings. He then enumerates the laws by which animals pass into one another, according to their degrees of knowledge or ignorance.

It will be seen, therefore, that Plato made animals to come into being by degradation from man, and according to him an Ape would be a degenerate Man, instead of Man being an improved Ape, as some of our modern philosophers would have us believe. I venture to affirm that there is quite as much evidence in favour of one view as of the other.

In thus commenting on the ever varying tendencies of science,* I need not say that nothing can be further from my intention than to discourage scientific study and research. I have been engaged in the pursuit of science during the greater part of my life, and I yield to none in my full recognition of the incalculable benefits accruing to mankind from the results of

* The Variations in Science, under the different heads of Astronomy, Geology, Anthropology, Egyptology, and Theology, are well set forth by the Rev. B. W. Savile, in a very erudite work entitled "The Truth of the Bible," in which the author deprecates the notion that Scripture, rightly understood, is opposed to the teachings of Science. He boldly asserts that the Book of Nature and the Book of Revelation equally lie open to our inspection, and that Religion has nothing to fear but everything to hope from the progress of real Science.

modern scientific discoveries. Science has conquered the elements; it has annihilated distance; it says to the Light—paint me that picture on that piece of glass; it says to Electricity—flash me that message with the speed of lightning to yon distant clime; it says to the Lightning itself— come thou down that rod and bury thyself in the earth! I hail these achievements as triumphs of human intellect, and I should as soon attempt to stop the progress of the avalanche which has become dislodged from the mountain top, as to try and bar the path of scientific progress and discovery. I am prepared to welcome light and knowledge from whatever quarter it may come, being fully convinced that all systems and theories irreconcilable with truth are built upon the sand and must ultimately be swept away. Nay more, I would not have scientists linger with complacency on the heights already attained, but with the confident assurance that fresh trophies are within their reach, and that fresh benefits are to be conferred on mankind, I would emblazon their scientific banner with the motto Excelsior, warning them, however, against the prevailing tendency to erect Science into an idol, to ignore the innate faculties of mankind, and to over-rule the dictates of common sense.

In conclusion, I desire to say that I entertain no preconceived hostility, no prejudice whatever, against Mr. Darwin and those who share his views, and I most certainly decline to be classed among those who would reject the doctrine of evolution simply from any fancied notion that its adoption is derogatory to man's position in the scheme of nature, for as Mr. Froude philosophically remarks "it is nothing to me how the Maker of me has been pleased to construct the organised substance which I call my body, It is mine, but it is not I. The νοῦς, the intellectual spirit, being an οὐσία—an essence—I believe to be an imperishable something which has been engendered in me from another source." Nor should I reject the evolution theory on the ground of any antagonism between it and the power of the Deity, for the same Power that planned the glorious temple of Nature, which has 'the earth for its emerald floor; its roof the sapphire firmament; the sun and stars its pendent lamps; its music the murmur of streams, the pealing thunder, and the everlasting roar of ocean,'—I say this same power could easily, during countless æons of geological ages, have caused us to pass through the probationary stages of ascidian, fish, reptile, monkey, and on to man, *if it had so willed it;* but as science has failed to show that it is so, I pin my faith to the story in the Grand Old Book, which tells us that man was created in the divine image, and I accept the tradition that Man sprang *as Man* direct from the hands of his God.

THE END.

Printed by Henry W. Stacy, Norwich.

www.ingramcontent.com/pod-product-compliance
Lightning Source LLC
Chambersburg PA
CBHW031727230426
43669CB00007B/272